THE HODGE THEORY OF
PROJECTIVE MANIFOLDS

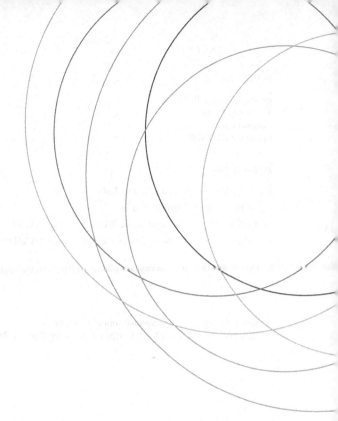

THE HODGE THEORY OF
PROJECTIVE MANIFOLDS

MARK ANDREA DE CATALDO

Stony Brook University, USA

Imperial College Press

ICP

Published by

Imperial College Press
57 Shelton Street
Covent Garden
London WC2H 9HE

Distributed by

World Scientific Publishing Co. Pte. Ltd.
5 Toh Tuck Link, Singapore 596224
USA office: 27 Warren Street, Suite 401-402, Hackensack, NJ 07601
UK office: 57 Shelton Street, Covent Garden, London WC2H 9HE

British Library Cataloguing-in-Publication Data
A catalogue record for this book is available from the British Library.

THE HODGE THOERY OF PROJECTIVE MANIFOLDS

ISBN-13 978-1-86094-800-8
ISBN-10 1-86094-800-6

Printed in Singapore.

Alla memoria di Meeyoung

Preface

After the mountain,
another mountain
(traditional Korean saying)

As my friend and colleague Luca Migliorini once wrote to me, the topology of algebraic varieties is a mystery and a miracle.

These lectures are an attempt to introduce the reader to the Hodge theory of algebraic varieties.

The geometric implications of Hodge theory for a compact oriented manifold become progressively richer and more beautiful as one specializes from Riemannian, to complex, to Kähler and finally to projective manifolds.

The structure of these lectures tries to reflect this fact.

I delivered eight one-hour lectures at the July 22 – July 27, 2003 Summer School on Hodge Theory at the Byeonsan Peninsula in South Korea. The present text is a somewhat expanded and detailed version of those lectures.

I would like to thank Professor JongHae Keum and Dr. Byungheup Jun for organizing the event, for editing a privately circulated version of these notes and for kindly agreeing to their publication.

I would like to thank Professor Jun-Muk Hwang and Professor Yongnam Lee for supporting the Summer School.

I would like to thank all of those who have attended the lectures for the warm atmosphere I have found that has made my stay a wonderful experience. In particular, I would like to thank Professor Dong-Kwan Shin for showing me with great humor some aspects of the Korean culture.

Beyond the choice of topics, exercises and exposition style, nothing in these written-up version of the lectures is original. The reader is assumed to have some familiarity with smooth and complex manifolds. These lectures are not self-contained and at times a remark or an exercise require knowledge of notions and facts which are not covered here. I see no harm in that, but rather as an encouragement to the reader to explore the subjects involved. The location of the exercises in the lectures is a suggestion to the reader to solve them before proceeding to the theoretical facts that follow.

The table of contents should be self-explanatory. The only exception is §8 where I discuss, in a simple example, a technique for studying the class map for homology classes on the fibers of a map and one for approximating a certain kind of primitive vectors. These techniques have been introduced in [de Cataldo and Migliorini, 2002] and [de Cataldo and Migliorini, 2005].

I would like to thank Fiammetta Battaglia, N. Hao, Gabriele LaNave, Jungool Lee and Luca Migliorini for suggesting improvements and correcting some mistakes. The remaining mistakes and the shortcomings in the exposition are mine.

During the preparation of these lecture notes, I have been supported by the following grants: N.S.F. Grant DMS 0202321, N.S.F. Grant DMS 0501020 and N.S.A. Grant MDA904-02-1-0100. The final version of the book was completed at I.A.S. (Princeton), under grant DMS 0111298.

Mark Andrea Antonio de Cataldo

Contents

Lecture 1

Calculus on smooth manifolds

We introduce some basic structures on a finite dimensional real vector space with a metric: the metric on the exterior algebra, orientations, the volume element and the star isomorphism. We introduce smooth differential forms, de Rham cohomology groups, state the de Rham Theorem and discuss Weil's sheaf-theoretic approach to this theorem. We briefly discuss Riemannian metrics, orientations on manifolds and integration of top forms on oriented manifolds.

References for this lecture are [Warner, 1971], [Bott and Tu, 1986] and [Demailly, 1996].

Manifolds are assumed to be *connected* and to satisfy the second axiom of countability. One of the advantages of the second assumption is that it implies the existence of partitions of unity, an important tool for the study of smooth manifolds.

1.1 The Euclidean structure on the exterior algebra

Let V be an m-dimensional real vector space with inner product $\langle\,,\,\rangle$, i.e. $\langle\,,\,\rangle : V \times V \longrightarrow \mathbb{R}$ is a symmetric, positive definite bilinear form on V.

Let

$$\Lambda(V) = \bigoplus_{0 \leq p \leq m} \Lambda^p(V)$$

be the exterior algebra associated with V.

1

If $\{e_1, \ldots, e_m\}$ is a basis for V, then the elements $e_I = e_{i_1} \wedge \ldots \wedge e_{i_p}$, where $I = (i_1, \ldots, i_p)$ ranges in the corresponding set of multi-indices with $1 \leq i_1 < \ldots < i_p \leq m$, form a basis for $\Lambda^p(V)$.

The elements of $\Lambda^p(V)$ can be seen as the alternating p-linear form on V^* as follows:

$$v_1 \wedge \ldots \wedge v_p = \sum_{\nu \in S_p} \epsilon(\nu) \, v_{\nu_1} \otimes \ldots \otimes v_{\nu_p},$$

where S_p is the symmetric group in p elements and $\epsilon(\nu)$ is the sign of the permutation ν.

When dealing with exterior algebras it is costumary to write summations such as $\sum_I u_I \, e_I$ or $\sum_{|I|=p} u_I \, e_I$ meaning that the summation is over the set of ordered multi-indices I as above and $u_I \in \mathbb{R}$.

There is a natural inner product on $\Lambda(V)$ defined by declaring any two distinct spaces $\Lambda^p(V)$ and $\Lambda^{p'}(V)$ mutually orthogonal and setting

$$\langle \, v_1 \wedge \ldots \wedge v_p, \, w_1 \wedge \ldots \wedge w_p \, \rangle := \det \| \, \langle v_j, w_k \rangle \, \|, \qquad v_j, w_k \in V. \quad (1.1)$$

If the basis $\{e_1, \ldots, e_m\}$ for V is orthonormal, then so is the corresponding one for $\Lambda(V)$.

Exercise 1.1.1. Let $V = V' \oplus V''$ be an orthogonal direct sum decomposition, $v', w' \in \Lambda^{p'}(V')$ and $v'', w'' \in \Lambda^{p''}(V'')$. Show that

$$\langle\langle v' \wedge v'', \, w' \wedge w'' \rangle\rangle = \langle\langle v', w' \rangle\rangle \, \langle\langle v'', w'' \rangle\rangle.$$

1.2   The star isomorphism on $\Lambda(V)$

By definition, $\Lambda^0(V) = \mathbb{R}$. The positive half-line $\mathbb{R}^+ \subseteq \Lambda^0(V)$ is defined without ambiguity.

The real vector space $\Lambda^m(V)$ is one-dimensional and $\Lambda^m(V) \setminus \{0\}$ has two connected components, i.e. two half-lines.

However, unlike the case of $\Lambda^0(V)$, there is no canonical way to distinguish either of them.

A choice is required. This choice gives rise to an isometry $\Lambda^0(V) \simeq \Lambda^m(V)$.

The $\star$ operator is the operator that naturally arises when we want to complete the picture with linear isometries $\Lambda^p(V) \simeq \Lambda^{m-p}(V)$.

Definition 1.2.1. (Orientations of V) The choice of a connected component $\Lambda^m(V)^+$ of $\Lambda^m(V) \setminus \{0\}$ is called an *orientation* on V.

Let V be oriented and $\{e_i\}$ be an ordered, orthonormal basis such that $e_1 \wedge \ldots \wedge e_m \in \Lambda^m(V)^+$. This element is uniquely defined since any two bases as above are related by an orthogonal matrix with determinant $+1$.

Definition 1.2.2. (The volume element) The vector

$$\boxed{dV := e_1 \wedge \ldots \wedge e_m \in \Lambda^m(V)^+} \tag{1.2}$$

is called *the volume element* associated with the oriented $(V, \langle \, , \, \rangle, \Lambda^m(V)^+)$.

Definition 1.2.3. (The $\star$ operator) The $\star$ *operator* is the unique linear isomorphism

$$\star : \Lambda(V) \simeq \Lambda(V)$$

defined by the properties

$$\star : \Lambda^p(V) \simeq \Lambda^{m-p}(V),$$

$$\boxed{u \wedge \star v = \langle u, v \rangle \, dV, \qquad \forall \, u, v \in \Lambda^p(V), \forall p.}$$

The $\star$ operator depends on the inner product *and* on the chosen orientation.

Let us check that the operator $\star$ exists and is unique.

Consider the non-degenerate pairing

$$\Lambda^p(V) \times \Lambda^{m-p}(V) \longrightarrow \mathbb{R}, \qquad (u, w) \longrightarrow (u \wedge w)/dV.$$

More explicitly, let $\{e_i\}$ be an oriented orthonormal basis, $dV = e_1 \wedge \ldots \wedge e_m$ be the volume element, $u = \sum_I u_I \, e_I$, $w = \sum_J w_J \, e_J$. Then

$$u \wedge w = \sum_I \epsilon(I, CI) \, u_I \, w_{CI} \, dV,$$

where, if I is an ordered set of indices, then CI is the complementary ordered set of indices and $\epsilon(I, CI)$ is the sign of the permutation (I, CI) of the ordered set $\{1, \ldots, m\}$.

Let $v = \sum_I v_I \, e_I$ and define

$$\boxed{\star v := \sum_I \epsilon(I, CI) \, v_I \, e_{CI}.} \tag{1.3}$$

We have that

$$u \wedge \star v = \sum_I u_I \, e_I \wedge \sum_{I'} \epsilon(I', CI') \, v_{I'} \, e_{CI'} =$$

$$= \sum_I u_I \, v_I \, \epsilon(I, CI) \, e_I \wedge e_{CI} = \langle u, v \rangle \, dV.$$

This shows the existence of $\star$, which seems to depend on the choice of the orthonormal basis.

Assume $\star'$ is another such operator. Then $u \wedge (\star - \star')(v) = (\langle u, v \rangle - \langle u, v \rangle) \, dV = 0$ for every u and v which implies that $\star = \star'$.

Exercise 1.2.4. Verify the following statements.

The $\star$ operator is a linear isometry.

Let $dV = e_1 \wedge \ldots \wedge e_m$ be the volume element for the fixed orientation.

$$\star(1) = dV, \quad \star(dV) = 1, \quad \star(e_1 \wedge \ldots \wedge e_p) = e_{p+1} \wedge \ldots \wedge e_m,$$

$$\star\star_{|\Lambda^p(V)} = (-1)^{p(m-p)} \, Id_{\Lambda^p(V)}, \tag{1.4}$$

$$\langle u, v \rangle = \star(v \wedge \star u) = \star(u \wedge \star v), \quad \forall \, u, v \in \Lambda^p(V), \quad \forall \, 0 \le p \le m.$$

1.3 The tangent and cotangent bundles of a smooth manifold

A smooth manifold M of dimension m comes equipped with natural smooth vector bundles.

Let $(U; x_1, \ldots, x_m)$ be a chart centered at a point $q \in M$.

- T_M the *tangent bundle* of M. The fiber $T_{M,q}$ can be identified with the linear span $\mathbb{R}\langle \partial_{x_1}, \ldots, \partial_{x_m} \rangle$.
- T_M^* the *cotangent bundle* of M. Let $\{dx_i\}$ be the dual basis of the basis $\{\partial_{x_i}\}$. The fiber $T_{M,q}^*$ can be identified with the span $\mathbb{R}\langle dx_1, \ldots, dx_m \rangle$.
- $\Lambda^p(T_M^*)$ the *p-th exterior bundle* of T_M^*. The fiber $\Lambda^p(T_M^*)_q = \Lambda^p(T_{M,q}^*)$ can be identified with the linear span $\mathbb{R}\langle \{dx_I\}_{|I|=p} \rangle$.
- $\Lambda(T_M^*) := \oplus_{p=0}^m \Lambda^p(T_M^*)$ the *exterior algebra bundle* of M.

As it is custumary and by slight abuse of notation, one can use the symbols ∂_{x_i} and dx_j to denote the corresponding sections over U of the tangent and cotangent bundles.

In this case, the collections $\{\partial_{x_i}\}$ and $\{dx_i\}$ form *local frames* for the bundles in question.

Exercise 1.3.1. Let M be a smooth manifold of dimension m. Take any definition in the literature for the tangent and cotangent bundles and verify the assertions that follow. Let $(U; x_1, \ldots, x_m)$ and $(U'; x_1', \ldots, x_m')$ be two local charts centered at $q \in M$. Show that the transition functions $\tau_{U'U}(x)$: $U \times \mathbb{R}^m \simeq U' \times \mathbb{R}^m$ ($\gamma_{U'U}(x)$, resp.) for the tangent bundle T_M (cotangent bundle T_M^*, resp.) are given, using the column notation for vectors in $\mathbb{R}^m$, by

$$\tau_{U'U}(x) = \left[J\left(\frac{x'}{x}\right)(x) \right]^t, \qquad \gamma_{U'U}(x) = \left[J\left(\frac{x'}{x}\right)(x) \right]^{-1},$$

where

$$J\left(\frac{x'}{x}\right)(x) = \begin{pmatrix} \frac{\partial x_1'}{\partial x_1}(x) & \cdots & \frac{\partial x_m'}{\partial x_1}(x) \\ \vdots & \cdots & \vdots \\ \frac{\partial x_1'}{\partial x_m}(x) & \cdots & \frac{\partial x_m'}{\partial x_m}(x) \end{pmatrix}.$$

Exercise 1.3.2. Determine the tangent bundle of the sphere $S^n \subseteq \mathbb{R}^{n+1}$ given by the equation $\sum_{j=1}^{n+1} x_j^2 = 1$. Let $S^{n-1} \subseteq S^n$ be an "equatorial" embedding, i.e. obtained by intersecting S^n with the hyperplane $x_{n+1} = 0$. Study the "normal bundle" exact sequence

$$0 \longrightarrow T_{S^{n-1}} \longrightarrow (T_{S^n})_{|S^{n-1}} \longrightarrow N_{S^{n-1},S^n} \longrightarrow 0.$$

1.4   The de Rham cohomology groups

Definition 1.4.1. (*p*-**forms**) The elements of the real vector space

$$\boxed{E^p(M) := C^\infty(M, \Lambda^p(T_M^*))}$$

of smooth real-valued sections of the vector bundle $\Lambda^p(T_M^*)$ are called (smooth differential) *p-forms* on M.

Let $d : E^p(M) \longrightarrow E^{p+1}(M)$ denote the exterior derivation of differential forms.

Exercise 1.4.2. A *p*-form u on M can be written on U as $u = \sum_{|I|=p} u_I dx_I$. Show that exterior derivation of forms is well-defined. More precisely, show that if one defines, locally on the chart $(U; x_1, \ldots, x_m)$,

$$du := \sum_j \frac{\partial u_I}{\partial x_j} dx_j \wedge dx_I,$$

then du is independent of the choice of coordinates.

Definition 1.4.3. (Complexes, cohomology of a complex) A *complex* is a sequence of maps of vector spaces

$$\cdots \longrightarrow V^{i-1} \xrightarrow{\delta^{i-1}} V^i \xrightarrow{\delta^i} V^{i+1} \longrightarrow \cdots, \quad i \in \mathbb{Z},$$

also denoted by $(V^\bullet, \delta)$, such that $\delta^i \circ \delta^{i-1} = 0$ for every index i, i.e. such that $\operatorname{Im} \delta^{i-1} \subseteq \operatorname{Ker} \delta^i$.

The vector spaces $H^i(V^\bullet, \delta) =: \operatorname{Ker} \delta^i / \operatorname{Im} \delta^{i-1}$ are called the *cohomology groups* of the complex.

A complex is said to be *exact at i* if $\operatorname{Im} \delta^{i-1} = \operatorname{Ker} \delta^i$, i.e. if $H^i(V^\bullet, \delta) = 0$, and *exact* if it is exact for every $i \in \mathbb{Z}$.

Exercise 1.4.4. (The de Rham complex) Show that $d^2 = 0$ so that we get the so-called *de Rham* complex of vector spaces,

$$\boxed{0 \longrightarrow E^0(M) \xrightarrow{d} E^1(M) \xrightarrow{d} \cdots \xrightarrow{d} E^{m-1}(M) \xrightarrow{d} E^m(M) \longrightarrow 0.}$$

$$(1.5)$$

Definition 1.4.5. (Closed/exact) A p-form u is said to be *closed* if $du = 0$ and is said to be *exact* if there exists $v \in E^{p-1}(M)$ such that $dv = u$.

Exercise 1.4.6. Let u and v be closed forms. Show that $u \wedge v$ is closed. Assume, in addition, that v is exact and show that $u \wedge v$ is exact.

Exercise 1.4.7. Let

$$u = (2x + y \cos xy) \, dx + (x \cos xy) \, dy$$

on $\mathbb{R}^2$. Show that u is exact. What is the integral of u along any closed curve in $\mathbb{R}^2$?

Exercise 1.4.8. Let

$$u = \frac{1}{2\pi} \frac{x \, dy - y \, dx}{x^2 + y^2}$$

on $\mathbb{R}^2 \setminus \{0\}$. Show that u is closed. Compute the integral of u over the unit circle S^1. Is u exact? Is $u_{|S^1}$ exact?

Exercise 1.4.9. (a) Prove that every closed 1-form on S^2 is exact.

(b) Let

$$u = \frac{x \, dy \wedge dz - y \, dx \wedge dz + z \, dx \wedge dy}{(x^2 + y^2 + z^2)^{3/2}}$$

on $\mathbb{R}^3 \setminus \{0\}$. Show that u is closed.

(c) Evaluate $\int_{S^2} u$. Conclude that u is not exact.

(d) Let

$$u = \frac{x_1 \, dx_1 \ldots + x_n \, dx_n}{(x_1^2 + \ldots + x_n^2)^{n/2}}$$

on $R^n \setminus \{0\}$. Show that $\star u$ is closed.

(e) Evaluate

$$\int_{S^{n-1}} \star u.$$

Is $\star u$ exact?

The de Rham complex (1.5) is *not* exact and its deviation from exactness is an important invariant of M and is measured by the so-called de Rham cohomology groups of M; see Theorem 1.4.12.

Definition 1.4.10. (The de Rham cohomology groups) The real *de Rham cohomology groups* $H^\bullet_{dR}(M, \mathbb{R})$ of M are the cohomology groups of the complex (1.5), i.e.

$$H^p_{dR}(M, \mathbb{R}) \simeq \frac{\text{closed } p\text{-forms on } M}{\text{exact } p\text{-forms on } M.}$$

The de Rham complex is *locally exact* on M by virtue of the important:

Theorem 1.4.11. (Poincaré Lemma) *Let $p > 0$. A closed p-form u on M is locally exact, i.e. for every $q \in M$ there exists an open neighborhood U of q and $v \in E^{p-1}(U)$ such that*

$$u_{|U} = dv.$$

Proof. See [Bott and Tu, 1986] §4, [Warner, 1971], §4.18.   □

The following result of de Rham's is fundamental. In what follows the algebra structures are given by the wedge and cup products, respectively.

Theorem 1.4.12. (The de Rham Theorem) *Let M be a not necessarily orientable smooth manifold. There is a canonical isomorphisms of $\mathbb{R}$-algebras*

$$H^\bullet_{dR}(M, \mathbb{R}) \simeq H^\bullet(M, \mathbb{R}).$$

Remark 1.4.13. Theorem 1.4.12 is obtained using integration over differentiable simplices and then using the various canonical identifications between real singular cohomology and real differentiable singular cohomology. Of course, one also has canonical isomorphisms with real Alexander-Spanier cohomology, sheaf cohomology with coefficients in the locally constant sheaf $\mathbb{R}_M$ and Čech cohomology with coefficients in $\mathbb{R}_M$. See [Warner, 1971], § 4.7, 4.17, 5.34 − 5.38, 5.43 − 5.45, and [Mumford, 1970], 5.23–5.28.

Remark 1.4.14. (Sheaf-theoretic de Rham Theorem) It is important to know that what above, and more, admits a sheaf-theoretic re-formulation, due to A. Weil. Here is a sketch of this re-formulation. Let $\mathcal{E}_M^p$ be the sheaf of germs of smooth p-forms on M, $\mathcal{E}_M := \mathcal{E}_M^0$ be the sheaf of germs of smooth functions on M, $\mathbb{R}_M$ be the sheaf of germs of locally constant functions on M. Due to the existence of partitions of unity, the sheaf $\mathcal{E}_M$ is what one calls a *fine* sheaf (see [Griffiths and Harris, 1978], p. 42). Any sheaf of $\mathcal{E}_M$-modules is fine. In particular, the sheaves $\mathcal{E}^p(M)$ are fine. Fine sheaves have trivial higher sheaf cohomology groups. The operator d is defined locally and gives rise to maps of sheaves $d : \mathcal{E}_M^p \longrightarrow \mathcal{E}_M^{p+1}$. In this context, the Poincaré Lemma 1.4.11 implies that the complex of sheaves

$$0 \longrightarrow \mathbb{R}_M \hookrightarrow \mathcal{E}_M^0 \xrightarrow{d} \mathcal{E}_M^1 \xrightarrow{d} \ldots \xrightarrow{d} \mathcal{E}^{m-1} \xrightarrow{d} \mathcal{E}_M^m \longrightarrow 0 \qquad (1.6)$$

is exact. In short, $\mathbb{R}_M \longrightarrow (\mathcal{E}^\bullet, d^\bullet)$ is a resolution of $\mathbb{R}_M$ by fine sheaves. The sheaf cohomology of a sheaf F on X is defined by considering a resolution $F \longrightarrow I^\bullet$ of F by injective sheaves and setting $H_{Sheaf}^p(X, F) := H^p(H^0(X, I^\bullet))$. Choosing another injective resolution gives canonically isomorphic sheaf cohomology groups. One can take a resolution of F by fine sheaves as well. Taking global sections in (1.6) and ignoring $\mathbb{R}_M$, we obtain the de Rham complex (1.5). It follows that there is a natural isomorphism between the sheaf cohomology of $\mathbb{R}_M$ and de Rham cohomology:

$$H_{Sheaf}^p(X, \mathbb{R}_M) \simeq H^p(E^\bullet, d) =: H_{dR}^p(X, \mathbb{R}).$$

The de Rham Theorem 1.4.12 follows from the natural identification of the standard singular cohomology with real coefficients with the sheaf cohomology of $\mathbb{R}_M$. See [Warner, 1971].

Surprisingly, while the left-hand side is a topological invariant, the right-hand side depends on the smooth structure. In other words: *the "number" of linearly independent smooth closed p-forms which are not exact is a topological invariant, independent of the smooth structure on M.* The advantage of this sheaf-theoretic approach to the de Rham Isomorphism Theorem is that it gives rise to a variety of isomorphisms in different contexts, as soon as one has Poincaré Lemma-type results. Another example is the sheaf-theoretic proof of the Dolbeault Isomorphism Theorem based on the Grothendieck-Dolbeault Lemma 3.7.4. See Remark 3.7.6.

Remark 1.4.15. It is important to know that the complexes (1.6) and (1.5) have counterparts in the theory of currents on a manifold:

$$0 \longrightarrow \mathbb{R}_M \hookrightarrow \mathcal{D}_M^0 \xrightarrow{d} \mathcal{D}_M^1 \xrightarrow{d} \ldots \xrightarrow{d} \mathcal{D}^{m-1} \xrightarrow{d} \mathcal{D}_M^m \longrightarrow 0, \qquad (1.7)$$

$$0 \longrightarrow D^0(M) \xrightarrow{d} D^1(M) \xrightarrow{d} \ldots \xrightarrow{d} D^{m-1}(M) \xrightarrow{d} D^m(M) \longrightarrow 0 \quad (1.8)$$

yielding results analogous to the ones of Remark 1.4.14. See [Griffiths and Harris, 1978], §3 for a quick introduction to currents.

The space of currents $D^p(X, \mathbb{R})$ is defined as the topological dual of the space of compactly supported smooth p-forms $E_c^{m-p}(M, \mathbb{R})$ endowed with the C^∞-topology.

The exterior derivation

$$d = D^p(M, \mathbb{R}) \longrightarrow D^{p+1}(M, \mathbb{R})$$

is defined by

$$d\,(T)(u) := (-1)^{p+1} T(du).$$

Currents have a local nature which allows to write them as differential forms with distribution coefficients. If M is oriented, then any differential p-form u is also a p-current via the assignment $v \mapsto \int_M u \wedge v$, $v \in E_c^{m-p}(M, \mathbb{R})$.

In this case, the complexes (1.6) and (1.7), and (1.5) and (1.8) are quasi-isomorphic, i.e. the natural injection induces isomorphisms at the level of cohomology.

By integration, a piecewise smooth oriented $(n-p)$-chain in M gives a p-current in $D^p(M)$.

Importantly, any closed analytic subvariety V of complex codimension d of a complex manifold X gives rise, via integration, to a $2d$-current $\int_V$ in $D^{2d}(X, \mathbb{R})$. By Stokes' Theorem for analytic varieties [Griffiths and Harris, 1978], p. 33, such a current is closed and therefore gives rise to a cohomology class $[\int_V]$ which coincides with the fundamental class $[V]$ of $V \subseteq X$.

Exercise 1.4.16. Let $X \to Y$ be the blowing up of a smooth complex surface at a point $y \in Y$. Show that

$$Rf_*\mathbb{R}_X \simeq Rf_*\mathcal{D}_X^\bullet \simeq f_*\mathcal{D}_X^\bullet.$$

Use the currents of integration

$$\int_X \quad \text{and} \quad \int_{f^{-1}(y)}$$

to construct an isomorphism

$$Rf_*\mathbb{R}_X \simeq \mathbb{R}_Y[0] \oplus \mathbb{R}_y[-2].$$

1.5 Riemannian metrics

A *Riemannian metric* g on a smooth manifold M is the datum of a smoothly-varying positive inner product $g(-,-)_q$ on the fibers of $T_{M,q}$ of the tangent bundle of M. This means that, using a chart $(U;x)$, the functions

$$g_{jk}(q) = g(\partial_{x_j}, \partial_{x_k})_q$$

are smooth on U.

A Riemannian metric induces an isomorphism of vector bundles $T_M \simeq_g T_M^*$ and we can naturally define a metric on T_M^*.

Exercise 1.5.1. (The dual metric) Verify the following assertions.

Let E be a real, rank r vector bundle on M, $(U;x)$ and $(V;y)$ be two local charts around $q \in M$, $\tau_U : E_{|U} \longrightarrow U \times \mathbb{R}^r \longleftarrow V \times \mathbb{R}^r : \tau_V$ be two trivializations of E on the charts and $\tau_{VU}(x) = \tau_V \circ \tau_U^{-1} : (U \cap V) \times \mathbb{R}^r \longrightarrow (V \cap U) \times \mathbb{R}^r$ be the transition functions which we view as an invertible $r \times r$ matrix of real-valued functions of $x \in U \cap V$.

A metric g on E is a smoothly-varying inner product on the fibers of E. It can be viewed as a symmetric bilinear map of vector bundles $E \times E \longrightarrow \mathbb{R}_M$. Any bilinear map $b : E \times E \longrightarrow \mathbb{R}_M$ can be written, on a chart (U, x), as a $r \times r$ matrix b_U subject to the relation

$$b_V = (\tau_{VU}^{-1})^t \, b_U \, \tau_{VU}^{-1}.$$

If b is non-degenerate, then its local representations b_U are non-degenerate and we can define a non-degenerate bilinear form on the dual vector bundle E, $b^* : E^* \times E^* \longrightarrow \mathbb{R}_M$ by setting

$$b_U^* := b_U^{-1},$$

where it is understood that we are using as bases to represent b^* the dual bases to the ones employed to represent b on U.

If b is symmetric, then so is b^*. If b is positive definite, then so is b^*. If $\{e_1, \ldots, e_m\}$ is an orthonormal frame for E, with respect to g, then the dual frame $\{e_1^*, \ldots, e_m^*\}$ is orthonormal for g^*.

Conclude that if (M, g) is a Riemannian manifold, then there is a unique metric g^* on T_M^* such that the isomorphism induced by the metric (every bilinear form $V \times V \longrightarrow \mathbb{R}$, induces a linear map $V \longrightarrow V^*$) $T_M \simeq_g T_M^*$ is an isometry on every fiber.

1.6   Partitions of unity

Recall that a (smooth) *partition of unity* on M is a collection $\{\rho_\alpha\}$ of non-negative smooth functions on M such that the sum $\sum_\alpha \rho_\alpha$ is locally finite on M and adds-up to the value 1. This means that for every $q \in M$ there is a neighborhood U of q in M such that $\rho_{\alpha|U} \equiv 0$ for all but finitely many indices α so that the sum $\sum_\alpha \rho_\alpha(q)$ is finite and adds to 1.

Definition 1.6.1. (Partition of unity) Let $\{U_\alpha\}_{\alpha \in A}$ be an open covering of M. A *partition of unity subordinate to the covering* $\{U_\alpha\}_{\alpha \in A}$ is a partition of unity $\{\rho_\alpha\}$ such that the support of each ρ_α is contained in U_α.

Note that, on a non-compact manifold, it is not possible in general to have a partition of unity subordinate to a given covering and such that the functions ρ_α have compact support, e.g. the covering of $\mathbb{R}$ given by the single open set $\mathbb{R}$.

Theorem 1.6.2. *(Existence of partitions of unity)* Let $\{U_\alpha\}_{\alpha \in A}$ be *an open covering of M. Then there are*

a) a partition of unity subordinate to $\{U_\alpha\}$ and

b) a partition of unity $\{\rho_j\}_{j \in J}$, where $J \neq A$ in general, such that (i) the support of every ρ_j is compact and (ii) for every index j there is an index α such that $supp\,(\rho_j) \subseteq U_\alpha$.

Proof. See [Warner, 1971], p. 10. $\square$

Exercise 1.6.3. Prove that every smooth manifold admits Riemannian metrics on it. (Hint: use partitions of unity. See [Bott and Tu, 1986], p. 42.)

1.7   Orientation and integration

Given any smooth manifold M, the space $\Lambda^m(T_M^*) \setminus M$, where M is embedded in the total space of the line bundle $\Lambda^m(T_M^*)$ as the zero section, has at most two connected components. The reader should verify this.

Definition 1.7.1. (Orientation) A smooth manifold M is said to be *orientable* if $\Lambda^m(T_M^*) \setminus M$ has two connected components, *non-orientable* otherwise.

If M is orientable, then the choice of a connected component of $\Lambda^m(T_M^*)\backslash M$ is called an *orientation* of M which is then said to be *oriented*.

Example 1.7.2. The *standard orientation* on $\mathbb{R}^m = \{(x_1, \ldots, x_m) \mid x_j \in \mathbb{R}\}$ is the one associated with

$$dx_1 \wedge \ldots \wedge dx_m.$$

Similarly, the torus $\mathbb{R}^m/\mathbb{Z}^m \simeq (S^1)^m$ is oriented using the form above descended to the torus.

Exercise 1.7.3. Let M be a differentiable manifold of dimension m. Show that the following three statements are equivalent. See [Warner, 1971], §4.2 and [Bott and Tu, 1986], §*I*.3.

(a) M is orientable.

(b) There is a collection of coordinate system $\{U^\alpha, x^\alpha\}$ such that

$$\det \left|\left| \frac{\partial x_i^\beta}{\partial x_j^\alpha}(x^\alpha) \right|\right| > 0, \qquad \text{on } U^\alpha \cap U^\beta.$$

(c) There is a nowhere vanishing m-form on M.

Remark 1.7.4. A collection of charts with the property specified in Exercise 1.7.3.(b) is called *orientation-preserving*. Such a collection determines uniquely an orientation preserving collection of coordinate systems containing it which is maximal with respect to inclusion. Such a maximal collection is called an *orientation preserving atlas* for the orientable M. Note that composing with the automorphism $x \to -x$ we obtain a different orientation preserving atlas. Given an orientation preserving atlas, we can orient M by choosing the connected component of $\Lambda^m(T_M^*) \backslash M$ containing the vectors $\tau_\alpha^*(dx_1 \wedge \ldots \wedge dx_m)(q)$, where $\tau_\alpha : U_\alpha \longrightarrow \mathbb{R}^m$ is a chart in the atlas, $q \in U_\alpha$ and $dx_1 \wedge \ldots \wedge dx_n$ is the canonical orientation of $\mathbb{R}^m$. Of course, one could choose the opposite one as well. However, if M is oriented, then we consider the orientation preserving atlas and coordinate systems that agree with the orientation in the sense mentioned above.

Exercise 1.7.5. Let $f : \mathbb{R}^{m+1} \longrightarrow \mathbb{R}$ be a smooth function such that $df(q) \neq 0$ for every $q \in \mathbb{R}^{m+1}$ such that $f(q) = 0$.

Show that the equation $f = 0$ defines a possibly empty collection of smooth and connected m-dimensional submanifolds of $\mathbb{R}^{m+1}$.

Let M be a connected component of the locus $(f = 0)$.

Let

$$T := \{ (q, x) \mid x \cdot df(q) = 0 \} \subseteq M \times R^{m+1}.$$

Show that

$$T_M \sim T$$

as smooth manifolds and as vector bundles over M.

Show that if M is unique, then $\mathbb{R}^{m+1} \setminus M$ has two connected components A^+ and A^- determined by the sign of the values of f so that $A := A^- \coprod M$ is an m-manifold with boundary.

Show that M carries a natural orientation associated with A : let $q \in M$, $\{v_1, \ldots, v_m\}$ be vectors in $T^*_{\mathbb{R}^{m+1}, q}$ such that $(df(q), v_1, \ldots, v_m)$ is an oriented basis for $\mathbb{R}^{m+1}$ at q. Check that one indeed gets an orientation for M by taking the vectors v restricted to $T^*_{M, q}$.

Compare the above with the notion of induced orientation on the boundary of a manifold with boundary in [Bott and Tu, 1986], p. 31. They coincide.

Compare the above with [Warner, 1971], Exercise 4.1 which asks to prove that a codimension one submanifold of $\mathbb{R}^{m+1}$ is orientable if and only if there is a nowhere-vanishing smooth normal vector field (i.e. a section ν of $T_{\mathbb{R}^{m+1}}$ defined over $M \subseteq \mathbb{R}^{m+1}$, smooth over M, such that $\nu(q) \perp T_{M, q}$ for every $q \in M$).

Compute the volume form associated with the natural orientation on M and the Riemannian metric on M induced by the Euclidean metric on $\mathbb{R}^{m+1}$. The answer is that it is the restriction to M of the contraction of $dx_1 \wedge \ldots \wedge dx_{m+1}$ with the oriented unit normal vector field along M. See [Warner, 1971], Exercise 20.a. See [Warner, 1971], p. 61 and [Demailly, 1996], p. 22 for the definition, properties and explicit form of the contraction operation.

Make all the above explicit in the case when $M = S^m$ is defined by the equation $f := \sum_{j=1}^{m+1} x_j^2 = 1$ in the Euclidean space $\mathbb{R}^{m+1}$.

Compute everything using the usual spherical coordinates of calculus books.

Solve Exercise 4.20.(b) of [Warner, 1971] which gives the volume form for surfaces $(x, y, \varphi(x, y))$ in $\mathbb{R}^3$.

Exercise 1.7.6. Prove that the Möbius strip, the Klein bottle and $\mathbb{RP}^2$ are non-orientable.

Prove that $\mathbb{RP}^m$ is orientable iff m is odd. (Hint: the antipodal map $S^m \longrightarrow S^m$ is orientation preserving iff m is odd.)

Let M be an oriented manifold of dimension m. In particular, M admits an orientation preserving atlas. It is using this atlas and partitions of unity that we can define the operation of integrating m-forms with compact support on M and verify Stokes' Theorem 1.7.8.

Let us discuss how integration is defined. See [Bott and Tu, 1986], §I.3 (complemented by Remark 1.7.4 above) or [Warner, 1971], §4.8.

Let σ be an m-form on M with compact support contained in a chart $(U; x)$ with trivialization $\tau_U : U \longrightarrow \mathbb{R}^m$.

The m-form $(\tau_U^{-1})^*\sigma = f(x)\,dx_1 \wedge \ldots \wedge dx_m$, for a unique compactly supported real-valued function $f(x)$ on $\mathbb{R}^m$.

Define

$$\int_M^U \sigma := \int_{\mathbb{R}^m} f(x)\,dx,$$

where dx denotes the Lebesgue measure on $\mathbb{R}^m$ and the left-hand side is the Riemann-Stijlties-Lebesgue integral of $f(x)$.

The trouble with this definition is that it depends on the chosen chart. If (V, y) is another such chart and $T : y \to x$ is the patching function, then

$$(\tau_V^{-1})^*\sigma = g(y)\,dy_1 \wedge \ldots \wedge dy_m = f(T(y))\,J(T)(y)\,dy_1 \wedge \ldots \wedge dy_m$$

so that

$$\int_M^V \sigma = \int_{\mathbb{R}^m} g(y)\,dy = \int_{\mathbb{R}^m} f(T(y))\,J(T)(y)\,dy.$$

On the other hand, the change of variable formula for the Riemann integral gives

$$\int_M^U \sigma := \int_{\mathbb{R}^m} f(x)\,dx = \int_{\mathbb{R}^m} f(T(y))\,|J(T)(y)|\,dy.$$

It follows that the two definitions agree iff $J(T)(y) > 0$ on the support of σ.

This suggests that we can define the integral of m-forms only in the presence of an orientation-preserving atlas.

Let ω be a smooth m-form with compact support on M.

Let $\{U_\alpha\}$ be an orientation preserving collection of coordinate systems (see Remark 1.7.4) with orientation preserving trivializations $\tau_\alpha : U_\alpha \simeq \mathbb{R}^m$ and $\{\rho_\alpha\}$ be a partition of unity subordinate to the covering $\{U_\alpha\}$.

The forms $\rho_\alpha \omega$ have compact support contained in U_α.

Define

$$\boxed{\int_M \omega = \sum_\alpha \int_{U_\alpha} \rho_\alpha \omega.} \tag{1.9}$$

One checks that it is well-defined by a simple partition of unity argument.

Changing the orientation simply changes the signs of the integrals.

Exercise 1.7.7. Check that $\int_M \omega$, as defined in (1.9), is well-defined, i.e. that the definition does not depend on the covering and partition of unity chosen.

Note that one can integrate compactly supported functions f on any Riemannian manifold so that, if the manifold is also oriented, then the integral coincides with the integral of the top-form $\star f$. See [Warner, 1971], p. 150.

Complex manifolds are always orientable and are usually oriented using a standard orientation; see Proposition 3.4.1. Using the standard orientation, if ω is the $(1,1)$-form associated with the Fubini-Study metric of the complex projective space $\mathbb{P}^n$ (see Exercise 5.1.4), then $\int_{\mathbb{P}^n} \omega^n = 1$.

Theorem 1.7.8. *(Stokes' Theorem (simple version)). Let M be an oriented manifold of dimension M and u be an $(m-1)$-form on M with compact support. Then*

$$\boxed{\int_M du = 0.}$$

Proof. See [Bott andt Tu, 1986], §*I*.3.5, [Warner, 1971], p. 148.   □

We have stated a weak version of Stokes' Theorem. See the references above for the complete version. See also [Warner, 1971], pp. 150-151 and Exercise 4.4 for its re-formulation as the Divergence Theorem.

If a Riemannian metric g on the oriented manifold M is given, then we have the notion of *Riemannian volume element* associated with (M, g) and with the orientation. It is the unique m-form dV on M such that, for every $q \in M$, dV_q is the volume element of Definition 1.2.2 for the dual metric g_q^* on $T_{M,q}^*$.

If the integral

$$\int_M dV$$

converges, then its value is positive and it is called the *volume* of the oriented Riemannian manifold.

Exercise 1.7.9. Compute the volume of the unit sphere $S^m \subseteq \mathbb{R}^{m+1}$ with respect to the metric induced by the Euclidean metric and the induced orientation.

Do the same for the tori $\mathbb{R}^m/\mathbb{Z}^m$, where the lattice $\mathbb{Z}^m \subseteq \mathbb{R}^m$ is generated by the vectors $(0, \ldots, r_j, \ldots, 0)$, $r_j \in \mathbb{R}^+$, $1 \le j \le m$.

Lecture 2

The Hodge theory of a smooth, oriented, compact Riemannian manifold

We define the inner product $\langle\langle\,,\,\rangle\rangle$, the adjoint d^*, the Laplacian, harmonic forms. We state the Hodge Orthogonal Decomposition Theorem for compact oriented Riemannian manifolds, deduce the Hodge Isomorphism Theorem and the Poincaré Duality Theorem.

Some references for this section are [Demailly, 1996], [Warner, 1971] and [Griffiths and Harris, 1978].

2.1 The adjoint of d : d^*

Let (M, g) be an oriented Riemannian manifold of dimension m.

By Exercise 1.5.1, where we set $(V, \langle,\rangle) = (T^*_{M,q}, g^*_q)$, the Riemannian metric g on M defines a smoothly varying inner product on the exterior algebra bundle $\Lambda(T^*_M)$.

By §1.2, the orientation on M gives rise to the $\star$ operator on the differential forms on M :

$$\boxed{\star\, :\, E^p(M) \longrightarrow E^{m-p}(M).} \tag{2.1}$$

In fact, the star operator is defined point-wise, using the metric and the orientation, on the exterior algebras $\Lambda(T^*_{M,q})$ and it extends to differential forms.

Remark 2.1.1. By looking at the degrees of the forms involved, it is clear that $\star$ and d do not commute. It is important to note that $\star$ does *not* preserve closed forms. See Exercise 2.2.2 and Lemma 2.3.2.

Definition 2.1.2. Define an inner product on the space of compactly supported p-forms on M by setting

$$\boxed{\langle\!\langle u,v \rangle\!\rangle := \int_M \langle u,v \rangle \, dV} = \int_M u \wedge \star v. \tag{2.2}$$

For the last equality, see Definition 1.2.3.

Once we have a metric, we may start talking about formal adjoints. For a more thorough discussion of this notion, see [Demailly, 1996], §2.6.

Definition 2.1.3. (Formal adjoints) Let $T : E^p(M) \longrightarrow E^{p'}(M)$ be a linear map. We say that a linear map

$$T^\star : E^{p'}(M) \longrightarrow E^p(M)$$

is the *formal adjoint* to T (with respect to the metric) if, for every compactly supported $u \in E^p(M)$ and $v \in E^{p'}(M)$:

$$\boxed{\langle\!\langle Tu,v \rangle\!\rangle = \langle\!\langle u,T^\star v \rangle\!\rangle.}$$

Definition 2.1.4. (Definition of $d^\star$) Define $d^\star : E^p(M) \longrightarrow E^{p-1}(M)$ as

$$\boxed{d^\star := (-1)^{m(p+1)+1} \star d \star.}$$

Proposition 2.1.5. *The operator $d^\star$ is the formal adjoint of d.*

Proof. We need to show that if $u_{p-1} \in E^{p-1}(M)$ and $v_p \in E^p(M)$ have compact support, then

$$\langle\!\langle du_{p-1},v_p \rangle\!\rangle = \langle\!\langle u_{p-1},d^\star v_p \rangle\!\rangle.$$

We have

$$\langle\!\langle du_{p-1},v_p \rangle\!\rangle = \int_M \langle du_{p-1},v_p \rangle \, dV = \int_M du_{p-1} \wedge \star v_p =$$

$$= \int_M d(u_{p-1} \wedge \star v_p) - (-1)^{p-1} u_{p-1} \wedge d \star v_p = \int_M (-1)^p u_{p-1} \wedge d \star v_p$$

$$= \int_M (-1)^p u_{p-1} \wedge (-1)^{(m-p+1)(p-1)} \star \star d \star v =$$

$$= \int_M u \wedge \star ((-1)^{m(p+1)+1} \star d \star v) = \langle\!\langle u,(-1)^{m(p+1)+1} \star d \star v \rangle\!\rangle \, dV$$

where the fourth equality follows from Stokes' Theorem 1.7.8, the fifth from (1.4) and the last one from (2.2) and simple $mod\,2$ congruences. $\qquad\square$

2.2 The Laplace-Beltrami operator of an oriented Riemannian manifold

Definition 2.2.1. (The Laplacian) The *Laplace-Beltrami operator,* or Laplacian, is defined as

$$\Delta : E^p(M) \longrightarrow E^p(M),$$

$$\boxed{\Delta := d^{\star} d + d\, d^{\star}.}$$

We have

$$\Delta = (-1)^{m(p+1)+1} d \star d \star + (-1)^{mp+1} \star d \star d.$$

While $\star$ is defined point-wise using the metric, $d^{\star}$ and Δ are defined locally (using d) and depend on the metric.

Exercise 2.2.2. Show that $\star\, \Delta = \Delta \star$. In particular, a form u is harmonic iff $\star\, u$ is harmonic.

The Laplacian is self-adjoint, i.e. it coincides with its adjoint and is an *elliptic* second order linear partial differential operator. See [Demailly, 1996], §4.14 and [Warner, 1971], 6.28, 6.35, Exercise 9, 16, 18 (the wave equation is not elliptic and there is no regularity theorem for it), 21 (elliptic operators for vector bundles).

Exercise 2.2.3. Verify that if (M, g) is the Euclidean space $\mathbb{R}^n$ with the standard metric and $u = \sum_{|I|=p} u_I \, dx_I$, then

$$\boxed{\Delta(u) = -\sum_I \sum_j \frac{\partial^2 u_I}{\partial x_j^2} \, dx_I.}$$

See [Demailly, http] §VI.3.12 and [Griffiths and Harris, 1978], p. 83.

2.3 Harmonic forms and the Hodge Isomorphism Theorem

Let (M, g) be a <u>compact</u> oriented Riemannian manifold.

Definition 2.3.1. (Harmonic forms) Define the space of *real harmonic p-forms* as

$$\mathcal{H}^p(M, \mathbb{R}) := \mathrm{Ker}\,(\,\Delta : E^p(M) \longrightarrow E^p(M)\,).$$

The space of harmonic forms depends on the metric. A form may be harmonic with respect to one metric and fail to be harmonic with respect to another metric.

Lemma 2.3.2. *A p-form $u \in E^p(M)$ is harmonic if and only if $du = 0$ and $d^\star u = 0$.*

Proof. It follows from the identity

$$\langle\langle \Delta u, u \rangle\rangle = \langle\langle dd^\star u + d^\star du, u \rangle\rangle = \langle\langle d^\star u, d^\star u \rangle\rangle + \langle\langle du, du \rangle\rangle = ||d^\star u||^2 + ||du||^2.$$

$\square$

Note that if M is *not* compact, then a form which is d-closed and $d^\star$-closed is harmonic by the definition of Δ. However, the converse is not true, e.g. the function x on $M = \mathbb{R}$. The argument above breaks down in handling not necessarily convergent integrals.

We now come to the main result concerning the Hodge theory of a compact oriented Riemannian compact manifold. The proof is based on the theory of elliptic differential operators. See [Warner, 1971], [Griffiths and Harris, 1978] for self-contained proofs.

Theorem 2.3.3. *(*The Hodge Orthogonal Decomposition Theorem*)*
Let (M, g) be a <u>compact</u> oriented Riemannian manifold.
  Then

$$\dim_{\mathbb{R}} \mathcal{H}^p(M, \mathbb{R}) < \infty$$

and we have a direct sum decomposition into $\langle\langle\,,\,\rangle\rangle$-orthogonal subspaces

$$E^p(M) = \mathcal{H}^p(M, \mathbb{R}) \overset{\perp}{\oplus} d(E^{p-1}(M)) \overset{\perp}{\oplus} d^\star(E^{p+1}(M)).$$

Exercise 2.3.4. Show that if u and v are m-forms on a compact oriented manifold M of dimension m such that $\int_M u = \int_M v$, then $(u - v)$ is exact.

Exercise 2.3.5. Show that the orthogonality of the direct sum decomposition is an easy consequence of Lemma 2.3.2.

Exercise 2.3.6. Show that if T and T^* are adjoint, then $\operatorname{Ker} T = (\operatorname{Im} T^*)^\perp$.

Corollary 2.3.7. *(The Hodge Isomorphism Theorem) Let (M, g) be a compact oriented Riemannian manifold. There is an isomorphism depending only on the metric g :*

$$\boxed{\mathcal{H}^p(M, \mathbb{R}) \simeq H^p_{dR}(M, \mathbb{R}).}$$

In particular, $\dim_\mathbb{R} H^p(M, \mathbb{R}) < \infty$.

Proof. By Lemma 2.3.2, d and d^* are adjoint.

By Exercise 2.3.6, we have the equality

$$\operatorname{Ker} d = \operatorname{Im} d^{\star\,\perp},$$

the right-hand side of which is, by virtue of Theorem 2.3.3,

$$\mathcal{H}^p(M, \mathbb{R}) \overset{\perp}{\oplus} dE^{p-1}(M).$$

The corollary follows. $\qquad\qquad\qquad\qquad\qquad\qquad\qquad\qquad\square$

The following is one way to think about Corollary 2.3.7. The inclusion $\operatorname{Im} d \subseteq \operatorname{Ker} d$ and its quotient $H^p_{dR}(M, \mathbb{R})$ are independent of the metric and canonically attached to the differentiable structure of the manifold M. The metric allows to distinguish a vector subspace, $\mathcal{H}^p(M, \mathbb{R})$, complementary to $\operatorname{Im} d$ in $\operatorname{Ker} d$ and hence isomorphic to $H^p_{dR}(M, \mathbb{R})$.

The following heuristic argument, which is taken from [Griffiths and Harris, 1978], p. 80, where the case of Dolbeault cohomology is treated, suggests why one may think that the results of Hodge theory should hold.

Let $\alpha = [u + dv] \in H^p_{dR}(M, \mathbb{R})$, where $u \in E^p(M)$ is a closed form representing α and $v \in E^{p-1}(M)$.

We have that $\|u\|$ *has minimal norm among all representatives of α iff* $d^*u = 0$, i.e., in view of Lemma 2.3.2, iff u is harmonic.

This means that the choice of an orientation and of a metric allows to distinguish a representative of any cohomology class in such a way that the norm is minimized and the representatives form a vector subspace of the closed forms.

The condition $d^\star u = 0$ is sufficient:

$$||u + dv||^2 \; = \; ||u||^2 + ||dv||^2 + 2\langle\langle u, dv\rangle\rangle \; \geq \; ||u||^2 + 2\langle\langle d^\star u, v\rangle\rangle \; = \; ||u||^2.$$

The condition $d^\star u = 0$ is necessary: if u has smallest norm, then

$$0 = \frac{d}{dt}\left(\,||u + t\,dv||^2\,\right)(0) = 2\,\langle\langle u, dv\rangle\rangle = 2\,\langle\langle d^\star u, v\rangle\rangle \quad \forall v \in E^{p-1}(M),$$

so that $d^\star u = 0$.

Let us emphasize again that the space $\mathcal{H}^p(M, \mathbb{R})$ and the identification with de Rham cohomology depend on the metric. Once the metric has been fixed, the content of Corollary 2.3.7 is that one has a unique harmonic representative for every de Rham cohomology class.

We are now in the position to give a proof of the Poincaré Duality isomorphism on an oriented compact manifold.

Theorem 2.3.8. (Poincaré Duality) *Let M be a compact oriented smooth manifold. The pairing*

$$H_{dR}^p(M, \mathbb{R}) \times H_{dR}^{m-p}(M, \mathbb{R}) \longrightarrow \mathbb{R}, \qquad (u, v) \longrightarrow \int_M u \wedge v$$

is non-degenerate, i.e. it induces an isomorphism (the Poincaré Duality Isomorphism)

$$\boxed{H_{dR}^{m-p}(M, \mathbb{R}) \simeq H_{dR}^p(M, \mathbb{R})^\vee.}$$

Proof. Note that the pairing, which is a priori defined at the level of forms, descends to de Rham cohomology by virtue of Stokes' Theorem (applied to $d(u \wedge v')$, $d(u' \wedge v)$ and to $du' \wedge dv'$ which is exact (check!)), i.e. if u and v are closed, then

$$\int_M (u + du') \wedge (v + dv') = \int_M u \wedge v.$$

Fix a Riemannian metric on M. Using the Hodge Isomorphism Theorem 2.3.7, we may use harmonic representatives. The key point is that u is harmonic if and only if $\star u$ is harmonic; see Exercise 2.2.2. We have the equality

$$\int_M u \wedge \star u = \int_M \langle u, u\rangle\, dV = ||u||^2.$$

This shows that given any non-zero harmonic form $u \in H_{dR}^p(M, \mathbb{R})$, there is a harmonic form, $\star u \in H_{dR}^{m-p}(M, \mathbb{R})$, such that it pairs non-trivially with u. This shows that the bilinear pairing induces injections $H_{dR}^p(M) \to H_{dR}^{m-p}(M)^\vee$ and, symmetrically, $H_{dR}^{m-p}(M) \to H^p(M)^\vee$. The result follows from the fact that injections of finite dimensional vector spaces of equal dimension are isomorphisms. □

Remark 2.3.9. The proof given above shows also that the $\star$ operator induces isomorphisms

$$\star : \mathcal{H}^p(M, \mathbb{R}) \simeq \mathcal{H}^{m-p}(M, \mathbb{R})$$

for any compact, smooth, oriented Riemannian manifold M. Via the Hodge Isomorphism Theorem 2.3.7, these induce isomorphisms

$$H_{dR}^p(M, \mathbb{R}) \simeq H_{dR}^{m-p}(M, \mathbb{R}).$$

These latter isomorphisms are not canonical. On the other hand, the Poincaré Duality isomorphisms depend only on the orientation. Changing the orientation, changes the sign of integrals and hence of the Poincaré Duality isomorphism.

Exercise 2.3.10. Let M be a smooth compact orientable manifold of dimension m, r be an even integer, $v \in H^r(M, \mathbb{Q})$, and define

$$L : H^\bullet(M, \mathbb{Q}) \longrightarrow H^{\bullet+r}(M, \mathbb{Q}), \qquad u \longrightarrow u \cup v.$$

Show that, in a suitable sense, L coincides with the dual map $L^\vee$.

State and prove the analogous statement for any r.

Lecture 3

Complex manifolds

We discuss various notions of conjugation on complex vector spaces, tangent and cotangent bundles on a complex manifold, the standard orientation of a complex manifold, the quasi complex structure, (p, q)-forms, d' and d'', Dolbeault and Bott-Chern cohomology.

3.1 Conjugations

Let us recall the following distinct notions of conjugation.

- There is of course the usual conjugation in $\mathbb{C} : \gamma \longrightarrow \overline{\gamma}$.
- Let V be a real vector space and $V_{\mathbb{C}} := V \otimes_{\mathbb{R}} \mathbb{C}$ be its complexification. There is the natural $\mathbb{R}$-linear isomorphism given by what we call *c-conjugation* (i.e. coming from the complexification)
  $$c : V_{\mathbb{C}} \longrightarrow V_{\mathbb{C}}, \qquad u \otimes \gamma \longrightarrow u \otimes \overline{\gamma}.$$
  We may denote $c(v)$ by $\overline{v}^c$
- Let V and V' be real vector spaces, $P : V_{\mathbb{C}} \longrightarrow V'_{\mathbb{C}}$ be complex linear. Define the *operational conjugation* the complex linear map
  $$\overline{P}^o : V_{\mathbb{C}} \longrightarrow V'_{\mathbb{C}}, \qquad \overline{P}^o(v) = \overline{P(\overline{v}^c)}. \tag{3.1}$$
  Note that the assignment $P \to \overline{P}^o$ defines a real linear isomorphism
  $$Hom_{\mathbb{C}}(V_{\mathbb{C}}, V'_{\mathbb{C}}) \simeq_{\mathbb{R}} Hom_{\mathbb{C}}(V_{\mathbb{C}}, V'_{\mathbb{C}})$$
  which is not complex linear. As the reader should check, $\overline{P}^c = \overline{P}^o$ in the space with conjugation $Hom_{\mathbb{C}}(V_{\mathbb{C}}, V'_{\mathbb{C}}) = Hom_{\mathbb{R}}(V, V') \otimes_{\mathbb{R}} \mathbb{C}$.
- Let W be a complex vector space. The *conjugate* $\overline{W}$ of W is the complex vector space such that $W = \overline{W}$ as real vector spaces, but such that the scalar multiplication in $\overline{W}$ is defined as
  $$\gamma \cdot w := \overline{\gamma}w$$

where, on the right-hand side we are using the given complex scalar multiplication on W. The identity map $Id_W : W \longrightarrow \overline{W}$ is a real linear isomorphism, but is not complex linear.

- Let $f : S \longrightarrow \mathbb{C}$ be a map of sets. Define $\overline{f}^r : S \longrightarrow \mathbb{C}$ as $\overline{f}^r(s) = \overline{f(s)}$.

As the reader should check, we have, for every $P \in Hom_\mathbb{C}(V_\mathbb{C}, V'_\mathbb{C})$

$$\boxed{\overline{P}^o(v) = \overline{P}^c(v) = \overline{P}^r(\overline{v}^c) = \overline{P(\overline{v}^c)}.} \tag{3.2}$$

3.2   Tangent bundles on a complex manifold

Let X be a complex manifold of dimension n, $x \in X$ and

$$(U \, ; \, z_1 = x_1 + i\, y_1, \ldots, z_n = x_n + i\, y_n)$$

be a holomorphic chart for X around x.

We have the following notions of "tangent bundles" for X. Depending on the context, the symbols ∂_{x_j}, dx_j, etc. will denote either vectors in the fibers at x of the corresponding bundles, or the corresponding local frames for those bundles.

- $T_X(\mathbb{R})$, the *real tangent bundle*. The fiber $T_{X,x}(\mathbb{R})$ has real rank $2n$ and it is the real span

$$\boxed{\mathbb{R}\langle\, \partial_{x_1}, \ldots, \partial_{x_n}, \partial_{y_1}, \ldots, \partial_{y_n} \,\rangle.}$$

- $T_X(\mathbb{C}) := T_X(\mathbb{R}) \otimes_\mathbb{R} \mathbb{C}$, the *complex tangent bundle*. The fiber $T_{X,x}(\mathbb{C}) := T_{X,x}(\mathbb{R}) \otimes_\mathbb{R} \mathbb{C}$ has complex rank $2n$ and is the complex span

$$\boxed{\mathbb{C}\langle\, \partial_{x_1}, \ldots, \partial_{x_n}, \partial_{y_1}, \ldots, \partial_{y_n} \,\rangle.}$$

Note that $T_X(\mathbb{R}) \subseteq T_X(\mathbb{C})$ via the natural real linear map $v \to v \otimes 1$. Set

$$\boxed{\partial_{z_j} := \tfrac{1}{2}(\partial_{x_j} - i\, \partial_{y_j}),} \qquad \boxed{\partial_{\overline{z}_k} := \tfrac{1}{2}(\partial_{x_j} + i\, \partial_{y_j}).} \tag{3.3}$$

Clearly, we have

$$\boxed{\partial_{x_j} = \partial_{z_j} + \partial_{\overline{z}_j},} \qquad \boxed{\partial_{y_j} = i\,(\partial_{z_j} - \partial_{\overline{z}_j}).} \tag{3.4}$$

We have $T_{X,x}(\mathbb{C}) = \mathbb{C}\langle\partial_{z_1}, \ldots, \partial_{z_n}, \partial_{\overline{z}_1}, \ldots, \partial_{\overline{z}_n}\rangle$.

In general, a smooth change of coordinates $(x, y) \longrightarrow (x', y')$ does not leave invariant the two subspaces $\mathbb{R}\langle\{\partial_{x_j}\}\rangle$ and $\mathbb{R}\langle\{\partial_{y_j}\}\rangle$ of $T_{X,x}(\mathbb{R})$.

A local linear change of coordinates $z'_j = \sum_k A_{jk} z_k$ (this is the key case to check when dealing with this kind of questions) produces a change of basis in $T_{X,x}(\mathbb{C})$:

$$\partial_{z'_j} = \sum_j (A^{-1})_{kj} \partial_{z_k}, \qquad \partial_{\bar{z}'_j} = \sum_j \overline{(A^{-1})_{kj}} \partial_{\bar{z}_k}.$$

The simple yet important consequence of this observation is that a holomorphic change of coordinates $z \longrightarrow z'$ fixes the two subspaces $\mathbb{C}\langle\{\partial_{z_j}\}\rangle \subseteq T_{X,x}(\mathbb{C})$ and $\mathbb{C}\langle\{\partial_{\bar{z}_j}\}\rangle \subseteq T_{X,x}(\mathbb{C})$. It follows that we may define the following complex smooth vector bundles:

- T'_X the *holomorphic tangent bundle* . The fiber

$$\boxed{T'_{X,x} = \mathbb{C}\langle \partial_{z_1}, \dots, \partial_{z_n} \rangle}$$

  has complex rank n. T'_X is a holomorphic vector bundle.
- T''_X the *anti-holomorphic tangent bundle*. The fiber

$$\boxed{T''_{X,x} = \mathbb{C}\langle \partial_{\bar{z}_1}, \dots, \partial_{\bar{z}_n} \rangle}$$

  has complex rank n. It is an anti-holomorphic vector bundle, i.e. it admits transition functions which are conjugate-holomorphic.

We have a canonical injection and a canonical internal direct sum decomposition into complex sub-bundles

$$\boxed{T_X(\mathbb{R}) \subseteq T_X(\mathbb{C}) = T'_X \oplus T''_X.} \tag{3.5}$$

Composing the injection with the projections we get canonical real isomorphisms

$$\boxed{T'_X \simeq_{\mathbb{R}} T_X(\mathbb{R}) \simeq_{\mathbb{R}} T''_X,} \tag{3.6}$$

$$\boxed{\partial_{z_j} \leftarrow \partial_{x_j} \rightarrow \partial_{\bar{z}_j},} \qquad \boxed{i\,\partial_{z_j} \leftarrow \partial_{y_j} \rightarrow -i\,\partial_{\bar{z}_j}} \tag{3.7}$$

The conjugation map $c : T_X(\mathbb{C}) \longrightarrow T_X(\mathbb{C})$, $c(v) = \bar{v}^c$, is a real linear isomorphism which is not complex linear and it induces a real linear isomorphism

$$\boxed{c : T'_X \simeq_{\mathbb{R}} T''_X,} \quad \boxed{c(\partial_{z_j}) = \partial_{\bar{z}_j},} \quad \boxed{c : T''_X \simeq_{\mathbb{R}} T'_X,} \quad \boxed{c(\partial_{\bar{z}_j}) = \partial_{z_j},}$$
$$\tag{3.8}$$

and complex linear isomorphisms

$$\boxed{c : T'_X \simeq_{\mathbb{C}} \overline{T''_X},} \qquad \boxed{c : T''_X \simeq_{\mathbb{C}} \overline{T'_X}.} \tag{3.9}$$

Exercise 3.2.1. (**Standard vector bundles on** $\mathbb{P}^n$) Verify the assertions that follow.

Let $\mathbb{P}^n$ be the complex projective space of dimension n. It is the set of equivalence classes of $(n+1)$-tuples of not-all-zero complex numbers $[x] = [x_0 : \ldots : x_n]$, where we identify two such tuples if they differ by a non-zero multiplicative constant. One endows $\mathbb{P}^n$ with the quotient topology stemming from the identification $\mathbb{P}^n = (\mathbb{C}^{n+1} \backslash \{0\})/\mathbb{C}^*$. $\mathbb{P}^n$ is a compact topological space. Consider the following open covering consisting of $n + 1$ open subsets of $\mathbb{P}^n : U^j = \{[x] \mid x_j \neq 0\}$, $0 \le j \le n$. We have homeomorphisms

$$\tau_j : U^j \longrightarrow \mathbb{C}^n, \qquad [x_0 : \ldots : x_n] \longrightarrow \left(z_0^j := \frac{x_0}{x_j}, \overset{j}{\ldots}, z_n^j := \frac{x_n}{x_j} \right).$$

There are biholomorphisms

$$\tau_k \circ \tau_j^{-1} : \mathbb{C}^n \setminus \{z_k^j = 0\} \longrightarrow \mathbb{C}^n \setminus \{z_j^k = 0\}$$

so that $\mathbb{P}^n$ is a complex manifold of dimension n. The transition functions are

$$\begin{cases} z_l^k = (z_k^j)^{-1} z_l^j \ l \neq j, \\ z_j^k = (z_k^j)^{-1}. \end{cases}$$

Let $a \in \mathbb{Z}$. Consider the nowhere vanishing holomorphic functions $(z_k^j)^{-a}$ on $U^j \cap U^k$. They define a holomorphic line bundle L_a on $\mathbb{P}^n$. For $a \ge 0$ the holomorphic sections of L_a are in natural bijection with the degree a polynomials in $\mathbb{C}[x_0, \ldots, x_n]$ so that the zero locus of a section coincides with the zero set of the corresponding polynomial (counting multiplicities).

The line bundle L_1 is called the *hyperplane bundle*.

The line bundles L_a, $a < 0$, have no holomorphic sections.

The line bundle L_{-1} is the tautological line bundle, i.e. the one that has as fiber over a point $[x] \in \mathbb{P}^n$ the line $\lambda(x_0, \ldots, x_n) \subseteq \mathbb{C}^{n+1}$.

One has $L_{a+b} \simeq L_a \otimes L_b$, and $L_a \simeq L_b$ iff $a = b$, even as topological complex vector bundles.

These are the only (topological complex) holomorphic line bundles on $\mathbb{P}^n$.

Use the atlas given above to compute the transition functions of $T'_{\mathbb{P}^n}$ and check that the anti-canonical line bundle $K^*_{\mathbb{P}^n} := \det T'_{\mathbb{P}^n} \simeq L_{n+1}$.

There is a non-splitting exact sequence (called the Euler sequence) of holomorphic vector bundles

$$0 \longrightarrow L_{-1} \longrightarrow L_0^{\oplus n+1} \longrightarrow T'_{\mathbb{P}^n} \otimes L_{-1} \longrightarrow 0.$$

The total Chern class of $T'_{\mathbb{P}^n}$ is $c(T'_{\mathbb{P}^n}) = (1 + c_1(L_1))^{n+1}$. See [Bott and Tu, 1986] for the definition of Chern classes and their basic properties.

3.3 Cotangent bundles on complex manifolds

Denote by

$$\{dx_1, \ldots, dx_n, dy_1, \ldots, dy_n\}$$

the basis dual to

$$\{\partial_{x_1}, \ldots, \partial_{x_n}, \partial_{y_1}, \ldots, \partial_{y_n}\},$$

by

$$\{dz_1, \ldots, dz_n\}$$

the basis dual to

$$\{\partial_{z_1}, \ldots, \partial_{z_n}\},$$

and by

$$\{d\bar{z}_1, \ldots, d\bar{z}_n\}$$

the basis dual to

$$\{\partial_{\bar{z}_1}, \ldots, \partial_{\bar{z}_n}\}.$$

Exercise 3.3.1. Verify the following identities.

$$\boxed{dz_j = dx_j + i\, dy_j,} \qquad \boxed{d\bar{z}_j = dx_j - i\, dy_j.} \qquad (3.10)$$

$$\boxed{dx_j = \tfrac{1}{2}(dz_j + d\bar{z}_j),} \qquad \boxed{dy_j = \tfrac{1}{2i}(dz_j - d\bar{z}_j).} \qquad (3.11)$$

We have the following vector bundles on X.

- $T_X^*(\mathbb{R})$, the *real cotangent bundle*, with fiber

$$\boxed{T_{X,x}^*(\mathbb{R}) \;=\; \mathbb{R}\langle\, dx_1, \ldots, dx_n, dy_1, \ldots, dy_n \,\rangle.}$$

- $T_X^*(\mathbb{C}) := (T_X(\mathbb{C}))^* = T_X^*(\mathbb{R}) \otimes \mathbb{C}$, the *complex cotangent bundle*, with fiber

$$\boxed{T_{X,x}^*(\mathbb{C}) \;=\; \mathbb{C}\langle\, dx_1, \ldots, dx_n, dy_1, \ldots, dy_n \,\rangle.}$$

- $T_X'^*$ the *holomorphic cotangent bundle* , with fiber

$$\boxed{T_{X,x}'^* \;=\; \mathbb{C}\langle\, dz_1, \ldots, dz_n \,\rangle.}$$

  It is a holomorphic vector bundle, dual to T_X'.

- $T_X''^*$ the *anti-holomorphic cotangent bundle* , with fiber

$$\boxed{T_{X,x}''^* \;=\; \mathbb{C}\langle\, d\bar{z}_1, \ldots, d\bar{z}_n \,\rangle.}$$

  It is an anti-holomorphic vector bundle, dual to T_X''.

We have a canonical injection and a canonical internal direct sum decomposition

$$\boxed{T_X^*(\mathbb{R}) \;\subseteq\; T_X^*(\mathbb{C}) \;=\; T_X'^* \oplus T_X''^*.} \tag{3.12}$$

Composing the injection with the projections we get canonical real isomorphisms

$$\boxed{T_X'^* \simeq_{\mathbb{R}} T_X^*(\mathbb{R}) \simeq_{\mathbb{R}} T_X''^*,} \tag{3.13}$$

$$\boxed{\tfrac{1}{2}\, dz_j \leftarrow dx_j \to \tfrac{1}{2}\, d\bar{z}_j,} \qquad\qquad \boxed{\tfrac{1}{2i}\, dz_j \leftarrow dy_j \to -\tfrac{1}{2i}\, d\bar{z}_j.} \tag{3.14}$$

The conjugation map $c : T_X^*(\mathbb{C}) \longrightarrow T_X^*(\mathbb{C})$ is a real linear isomorphism, which is not complex linear. It induces a real linear isomorphism

$$\boxed{c : T_X'^* \simeq_{\mathbb{R}} T_X''^*,} \quad \boxed{c(dz_j) = d\bar{z}_j,} \quad \boxed{c : T_X''^* \simeq_{\mathbb{R}} T_X'^*,} \quad \boxed{c(d\bar{z}_j) = dz_j,}$$
$$\tag{3.15}$$

and complex linear isomorphisms

$$\boxed{c : T_X'^* \simeq_{\mathbb{C}} \overline{T_X''^*},} \qquad\qquad \boxed{c : T_X''^* \simeq_{\mathbb{C}} \overline{T_X'^*}.} \tag{3.16}$$

Let $f(x_1, y_1, \ldots, x_n, y_n) = u(x_1, y_1, \ldots, x_n, y_n) + i\, v(x_1, y_1, \ldots, x_n, y_n)$ be a smooth complex-valued function in a neighborhood of x. We have

$$\boxed{df \;=\; du + i\, dv \;=\; \sum_j \frac{\partial f}{\partial z_j} dz_j + \sum_j \frac{\partial f}{\partial \bar{z}_j} d\bar{z}_j.} \tag{3.17}$$

Exercise 3.3.2. Verify the equality (3.17).

3.4 The standard orientation of a complex manifold

Proposition 3.4.1. *A complex manifold X admits a canonical orientation.*

Proof. By Exercise 1.7.3 it is enough to exhibit an atlas where the transition functions have positive determinant Jacobian. The holomorphic atlas is one such. In fact, the determinant Jacobian with respect to the change of coordinates $(x_1, y_1, \ldots, x_n, y_n) \longrightarrow (x_1', y_1', \ldots, x_n', y_n')$ is positive being the square of the absolute value of the determinant Jacobian of the holomorphic change of coordinates $(z_1, \ldots, z_n) \longrightarrow (z_1', \ldots, z_n')$. $\qquad\square$

Clearly, in what above, by "canonical," we mean naturally induced by the chosen orientation on the complex line (real plane).

Here is a more direct proof of Proposition 3.4.1.

Let $(U; z)$ be a chart in the holomorphic atlas of X, $x \in U$.

The real $2n$-form

$$\boxed{o_U := dx_1 \wedge dy_1 \wedge \ldots \wedge dx_n \wedge dy_n} \qquad (3.18)$$

is nowhere vanishing on U and defines an orientation for U.

One checks that

$$o_U = (i/2)^n \, dz_1 \wedge d\bar{z}_1 \wedge \ldots \wedge dz_n \wedge d\bar{z}_n$$

and that

$$o_U = (i/2)^n \, (-1)^{\frac{(n-1)n}{2}} \, dz_1 \wedge \ldots \wedge dz_n \wedge d\bar{z}_1 \wedge \ldots \wedge d\bar{z}_n.$$

Let $(U'; z')$ be another chart around x.

We have

$$J(z(x)) := \det \left\| \frac{\partial z_j'}{\partial z_k}(z(x)) \right\| > 0.$$

Since

$$dz_1' \wedge \ldots \wedge dz_n' \wedge d\bar{z}_1' \wedge \ldots \wedge d\bar{z}_n' = |J(z(x))|^2 \, dz_1 \wedge \ldots \wedge dz_n \wedge d\bar{z}_1 \wedge \ldots \wedge d\bar{z}_n,$$

we have that

$$o_{U'} = |J(z(x))|^2 \, o_U.$$

It follows that if we use a covering of X by means of holomorphic charts, we can glue the forms o_U using a partition of unity subordinate to the covering and obtain a nowhere vanishing real $2n$-form o which orients X independently of the covering chosen within the holomorphic atlas.

This is the so-called *standard orientation* of X; at every point $x \in X$ it is determined by the vector (3.18).

3.5 The quasi complex structure

The holomorphic tangent bundle T'_X of a complex manifold X admits the complex linear automorphism given by multiplication by i.

Via the isomorphism (3.6) we get an automorphism J of the real tangent bundle $T_X(\mathbb{R})$ such that $J^2 = -Id_{T_X(\mathbb{R})}$.

The same is true for T'^*_X using the dual map J^*.

Exercise 3.5.1. Show that, using a local chart $(U; z)$:

$$J(\partial_{x_j}) = \partial_{y_j}, \qquad J(\partial_{y_j}) = -\partial_{x_j},$$

$$J^*(dx_j) = -dy_j, \qquad J^*(dy_j) = dx_j.$$

The various properties of tangent and cotangent bundles, e.g. (3.5), (3.12), (3.6), (3.13) etc. can be seen using J via the eigenspace decomposition of $T_X(\mathbb{C})$ with respect to $J \otimes Id_{\mathbb{C}}$.

The formalism of quasi complex structures allows to start with a complex vector space V and end up with a display like (3.5):

$$V_{\mathbb{R}} \subseteq V_{\mathbb{R}} \otimes_{\mathbb{R}} \mathbb{C} = V' \oplus V''$$

where $V_{\mathbb{R}}$ is the real vector space underlying V, $V' \simeq_{\mathbb{C}} V$ and $V'' \simeq_{\mathbb{C}} \overline{V'}$ and $V_{\mathbb{R}} \otimes_{\mathbb{R}} \mathbb{C}$ has the conjugation operation.

A second equivalent point of view is detailed in Exercise 3.5.6 and starts with a V as above and considers $V \oplus \overline{V}$ instead.

We mention this equivalence in view of the use of Hermitean metrics on T'_X.

These metrics are defined as special tensors in $T'^*_X \otimes_{\mathbb{C}} \overline{T'^*_X}$.

However, in view of the fact that it is convenient to conjugate them, it may be preferable to view them as tensors in the space with conjugation $T^*_X(\mathbb{C}) \otimes_{\mathbb{C}} T^*_X(\mathbb{C})$: using the canonical isomorphism (3.16) we can view the tensor h as an element of $T'^*_X \otimes_{\mathbb{C}} T''^*_X \subseteq T^*_X(\mathbb{C}) \otimes_{\mathbb{C}} T^*_X(\mathbb{C})$.

This is also convenient in view of the use of the real alternating form associated with a Hermitean metric which can then be viewed as a real element of $\Lambda^2_{\mathbb{C}}(T^*_X(\mathbb{C}))$.

Definition 3.5.2. (Quasi complex structure) A *quasi complex structure* on a real vector space $V_{\mathbb{R}}$ of finite even dimension $2n$ is a $\mathbb{R}$-linear automorphism

$$J_{\mathbb{R}} : V_{\mathbb{R}} \simeq_{\mathbb{R}} V_{\mathbb{R}}, \qquad J^2 = -Id_{V_{\mathbb{R}}}.$$

Exercise 3.5.3. Show that giving a quasi complex structure $(V_{\mathbb{R}}, J_{\mathbb{R}})$ as in Definition 3.5.2 is equivalent to endowing $V_{\mathbb{R}}$ with a structure of complex vector space of dimension n. (Hints: in one direction define $i v := J_{\mathbb{R}}(v)$; in the other define $J_{\mathbb{R}}$ as multiplication by i.)

Let $(V_{\mathbb{R}}, J_{\mathbb{R}})$ be a quasi complex structure.

Let $V_{\mathbb{C}} := V_{\mathbb{R}} \otimes_{\mathbb{R}} \mathbb{C}$ and $J_{\mathbb{C}} = J_{\mathbb{R}} \otimes Id_{\mathbb{C}} : V_{\mathbb{C}} \simeq_{\mathbb{C}} V_{\mathbb{C}}$ be the complexification of $J_{\mathbb{R}}$.

The automorphism $J_{\mathbb{C}}$ of $V_{\mathbb{C}}$ has eigenvalues i and $-i$.

There are a natural inclusion and a natural internal direct sum decomposition

$$\boxed{V_{\mathbb{R}} \subseteq V_{\mathbb{C}} = V' \oplus V''}$$

where

- the subspace $V_{\mathbb{R}} \subseteq V_{\mathbb{C}}$ is the fixed locus of the conjugation map associated with the complexification,
- V' and V'' are the $J_{\mathbb{C}}$-eigenspaces corresponding to the eigenvalues i and $-i$, respectively,
- since $J_{\mathbb{C}}$ is real, i.e. it fixes $V_{\mathbb{R}} \subseteq V_{\mathbb{C}}$, $J_{\mathbb{C}}$ commutes with the natural conjugation map and V' and V'' are exchanged by this conjugation map,
- there are natural $\mathbb{R}$-linear isomorphisms coming from the inclusion and the projections to the direct summands

$$\boxed{V' \simeq_{\mathbb{R}} V_{\mathbb{R}} \simeq_{\mathbb{R}} V''}$$

and complex linear isomorphisms

$$\boxed{V' \simeq_{\mathbb{C}} \overline{V''},} \qquad \boxed{V'' \simeq_{\mathbb{C}} \overline{V'},}$$

- the complex vector space defined by the complex structure (see Exercise 3.5.3) is $\mathbb{C}$-linearly isomorphic to V'.

Exercise 3.5.4. Verify all the assertions made above.

The same considerations are true for the quasi complex structure $(V_{\mathbb{R}}^*, J_{\mathbb{R}}^*)$. We have

$$\boxed{V_{\mathbb{R}}^* \subseteq V_{\mathbb{C}}^* = V'^* \oplus V''^*,}$$

$$\boxed{V'^* \simeq_{\mathbb{R}} V_{\mathbb{R}}^* \simeq_{\mathbb{R}} V''^*,}$$

$$\boxed{V'^* \simeq_{\mathbb{C}} \overline{V''^*},} \qquad \boxed{V''^* \simeq_{\mathbb{C}} \overline{V'^*}.}$$

In addition to the natural conjugation map stemming from the complexification, $V_{\mathbb{C}}^*$ comes with the operational conjugation map, for its elements are in $Hom_{\mathbb{C}}(V_{\mathbb{C}}, \mathbb{R} \otimes_{\mathbb{R}} \mathbb{C})$. However, these two operations coincide; see (3.2).

Exercise 3.5.5. Verify all the assertions made above.

Exercise 3.5.6. Let $\widetilde{W} := W \oplus \overline{W}$ and $\iota : \widetilde{W} \longrightarrow \widetilde{W}$ be the involution exchanging the summands. Let $W_0 \subseteq \widetilde{W}$ be the real subspace fixed by ι.

Show that there is a natural quasi complex structure on W_0 and that we get the following structure

$$W_0 \subseteq W_{0\mathbb{C}} = W' \oplus W''$$

endowed with the natural conjugation c coming from the complexification.

Show that there is a natural isomorphism $l : W_{0\mathbb{C}} \simeq_{\mathbb{C}} \widetilde{W}$, mapping $W' \simeq_{\mathbb{C}} W$ and $W'' \simeq_{\mathbb{C}} \overline{W}$, which is compatible with the conjugation and involution, i.e. $\iota \circ l = l \circ c$.

Verify that

$$W' = l^{-1}(W) = \{ (w, w) \otimes 1 - (iw, iw) \otimes i \mid w \in W \}$$

and

$$W'' = l^{-1}(\overline{W}) = \{ (w, w) \otimes 1 + (iw, iw) \otimes i \mid w \in W \}.$$

The complex vector space

$$F = Hom_{\mathbb{R}}(W, \mathbb{C}) = W^* \oplus \overline{W}^*$$

admits a conjugation-type map $\overline{(f, \overline{g})} := (g, \overline{f})$ (here $\overline{g}(w) := \overline{g(w)}$) with fixed locus

$$F_0 = \{ (f, \overline{f}) \mid f \in W^* \}.$$

Show that the map

$$m : F_0 \otimes_{\mathbb{R}} \mathbb{C} \longrightarrow F,$$

$$m((f, \overline{f}) \otimes i) := (i\,f, \overline{i\,f})$$

is an isomorphism compatible with the conjugations. Verify that

$$W'^* = m^{-1}(W^*) = \{\, (f, \overline{f}) \otimes 1 - (i\,f, \overline{i\,f}) \otimes i \mid f \in W^* \,\}$$

and

$$W''^* = m^{-1}(\overline{W}^*) = \{\, (f, \overline{f}) \otimes 1 + (i\,f, \overline{i\,f}) \otimes i \mid f \in W^* \,\}.$$

3.6 Complex-valued forms

Definition 3.6.1. Let M be a smooth manifold. Define the *complex-valued smooth p-forms* as

$$\boxed{A^p(M) := E^p(M) \otimes_{\mathbb{R}} \mathbb{C} \simeq C^\infty(M, T_M(\mathbb{C})).}$$

The notion of exterior differentiation extends to complex-valued differential forms:

$$\boxed{d : A^p(M) \longrightarrow A^{p+1}(M).}$$

Remark 3.6.2. If M is a smooth manifold, then we have the complex-valued version of the de Rham Isomorphism. If M is a smooth, oriented, compact manifold, then we have the complex-valued versions of the Hodge theory statements of §2.3. No new idea is necessary for these purposes. In the remaining part of these lectures, instead, we are going to discuss the aspects of Hodge theory which are specific to complex, Kähler and projective manifolds.

Let X be a complex manifold of dimension n, $x \in X$, (p, q) be a pair of non-negative integers and define complex vector spaces

$$\boxed{\Lambda^{p,q}(T^*_{X,x}) := \Lambda^p(T'^*_{X,x}) \otimes \Lambda^q(T''^*_{X,x}) \subseteq \Lambda^{p+q}_{\mathbb{C}}(T^*_{X,x}(\mathbb{C})).} \qquad (3.19)$$

There is a canonical internal direct sum decomposition of complex vector spaces

$$\Lambda^l_{\mathbb{C}}(T^*_{X,x}(\mathbb{C})) \;=\; \bigoplus_{p+q=l} \Lambda^{p,q}(T^*_{X,x}).$$

Exercise 3.6.3. Verify that a holomorphic change of coordinates leaves this decomposition invariant. In particular, we can define smooth complex vector bundles $\Lambda^{p,q}(T^*_X)$ and obtain an internal direct sum decomposition of smooth complex vector bundles

$$\Lambda^l_{\mathbb{C}}(T^*_X(\mathbb{C})) \;=\; \bigoplus_{p+q=l} \Lambda^{p,q}(T^*_X). \tag{3.20}$$

Definition 3.6.4. ((p,q)-**forms**) The *space of (p,q)-forms* on X

$$A^{p,q}(X) \;:=\; C^{\infty}(\,X\,,\,\Lambda^{p,q}(T^*_X)\,)$$

is the complex vector space of smooth sections of the smooth complex vector bundle $\Lambda^{p,q}(T^*_X)$.

Exercise 3.6.5. Verify the following statements.

$A^{p,q}(X)$ is the complex vector space of smooth $(p+q)$-forms u which can be written, locally on U, with respect to any holomorphic chart $(U;z)$, as

$$u = \sum_{|I|=p,|J|=q} u_{IJ}(z)\, dz_I \wedge d\bar{z}_J.$$

There is a canonical direct sum decomposition,

$$A^l(X) \;=\; \bigoplus_{p+q=l} A^{p,q}(X)$$

and

$$d\,(A^{p,q}) \;\subseteq\; A^{p+1,q}(X) \oplus A^{p,q+1}(X).$$

Let $l = p + q$ and consider the natural projections

$$\pi^{p,q} : A^l(X) \longrightarrow A^{p,q}(X) \subseteq A^l(X).$$

Definition 3.6.6. Define operators

$$d' : A^{p,q}(X) \longrightarrow A^{p+1,q}(X), \qquad d'' : A^{p,q}(X) \longrightarrow A^{p,q+1}(X)$$

$$\boxed{d' = \pi^{p+1,q} \circ d,} \qquad\qquad \boxed{d'' = \pi^{p,q+1} \circ d.}$$

Note that

$$\boxed{d = d' + d'',} \qquad \boxed{d''^2 = 0 = d'^2,} \qquad \boxed{d'd'' = -d''d'.} \qquad (3.21)$$

Exercise 3.6.7. Verify that if, in local coordinates,

$$u = \sum_{|I|=p,|J|=q} u_{IJ}\, dz_I \wedge d\bar{z}_J \in A^{p,q}(X),$$

then

$$d'u = \sum_{I,J} \sum_{j} \frac{\partial u_{IJ}}{\partial z_j}\, dz_j \wedge dz_I \wedge d\bar{z}_J,$$

$$d''u = \sum_{I,J} \sum_{j} \frac{\partial u_{IJ}}{\partial \bar{z}_j}\, d\bar{z}_j \wedge dz_I \wedge d\bar{z}_J.$$

Exercise 3.6.8. Show that

$$\boxed{\overline{d'} = d'',} \qquad \boxed{\overline{d''} = d',}$$

where the conjugation symbol denotes the operational conjugation defined in §3.1.

3.7   Dolbeault and Bott-Chern cohomology

In this section, we look at new cohomology groups stemming from the complex structure, the Dolbeault cohomology groups $H_{d''}^{p,q}(X)$ and the Bott-Chern cohomology groups $H_{BC}^{p,q}(X)$.

The former ones appear in the complex-manifold-analogues of the results of §2.3, which should not be confused with the complex-valued-analogues of Remark 3.6.2.

The latter ones will be used to show the important fact that on a compact Kähler manifold the Hodge Decomposition is independent of the Kähler metric used to obtain it.

Definition 3.7.1. (The Dolbeault complex) Fix p and q. The *Dolbeault complex* of X is the complex of vector spaces

$$0 \longrightarrow A^{p,0}(X) \xrightarrow{d''} A^{p,1}(X) \xrightarrow{d''} \ldots \xrightarrow{d''} A^{p,n-1}(X) \xrightarrow{d''} A^{p,n}(X) \longrightarrow 0.$$
$$(3.22)$$

We have the analogous complex

$$0 \longrightarrow A^{0,q}(X) \xrightarrow{d'} A^{1,q}(X) \xrightarrow{d'} \ldots \xrightarrow{d'} A^{n-1,q}(X) \xrightarrow{d'} A^{n,q}(X) \longrightarrow 0.$$
$$(3.23)$$

Definition 3.7.2. (Dolbeault cohomology) The *Dolbeault cohomology groups* are the cohomology groups of the complex (3.22)

$$H_{d''}^{p,q}(X) := \frac{\operatorname{Ker} d'' : A^{p,q}(X) \longrightarrow A^{p,q+1}(X)}{\operatorname{Im} d'' : A^{p,q-1}(X) \longrightarrow A^{p,q}(X)}$$

Define also

$$H_{d'}^{p,q}(X) := \frac{\operatorname{Ker} d' : A^{p,q}(X) \longrightarrow A^{p+1,q}(X)}{\operatorname{Im} d' : A^{p-1,q}(X) \longrightarrow A^{p,q}(X)}.$$

Exercise 3.7.3. Show that conjugation induces canonical complex linear isomorphisms

$$H_{d''}^{p,q}(X) \simeq_{\mathbb{C}} \overline{H_{d'}^{q,p}(X)}.$$

Theorem 3.7.4. *(Grothendieck-Dolbeault Lemma) Let $q > 0$. Let X be a complex manifold and $u \in A^{p,q}(X)$ be such that $d''u = 0$. Then, for every point $x \in X$, there is an open neighborhood U of x in X and a form $v \in A^{p,q-1}(U)$ such that*

$$u_{|U} = d''v.$$

Proof. See [Griffiths and Harris, 1978], p. 25. □

Exercise 3.7.5. (Grothendieck-Dolbeault Lemma for d'.) State and prove the result analogous to Theorem 3.7.4 for d'.

Remark 3.7.6. This remark is the Dolbeault counterpart of the Weil-de Rham isomorphism Theorem of Remark 1.4.14. We have the fine sheaves $\mathcal{A}_X^{p,q}$ of germs of local (p,q)-forms, the sheaf Ω_X^p of germs of holomorphic p-forms on X, i.e. of the form $\sum_{|I|=p} u_I dz_I$, u_I holomorphic functions. The Grothendieck-Dolbeault Lemma 3.7.4 implies that $\Omega_X^p \longrightarrow (\mathcal{A}_X^{p,\bullet}, d'')$ is a resolution of Ω_X^p by fine sheaves.

We get the canonical *Dolbeault Isomorphisms*

$$\boxed{H^q(X, \Omega_X^p) \simeq H_{d''}^{p,q}(X).}$$

The d'-version of the Grothendieck-Dolbeault Lemma gives isomorphisms

$$\boxed{H^p(X, \overline{\Omega_X^q}) \simeq H_{d'}^{p,q}(X)}$$

where $\overline{\Omega_X^q}$ is the sheaf of germs of anti-holomorphic q-forms on X, i.e. of the form $\sum_{|I|=q} u_I d\bar{z}_I$, u_I anti-holomorphic functions.

Definition 3.7.7. (Bott-Chern cohomology) The *Bott-Chern cohomology groups* of X are defined as the quotient spaces

$$\boxed{H_{BC}^{p,q}(X) := \frac{A^{p,q}(X) \cap \operatorname{Ker} d}{d'd''(A^{p-1,q-1}(X))}.}$$

One can prove that if X is compact, then $\dim H_{BC}^{p,q}(X) < \infty$.

Exercise 3.7.8. Verify that there are natural maps

$$H_{BC}^{p,q}(X) \longrightarrow H_{d''}^{p,q}(X), \ H_{BC}^{p,q}(X) \longrightarrow H_{d'}^{p,q}(X), \ H_{BC}^{p,q}(X) \longrightarrow H_{dR}^{p+q}(X, \mathbb{C})$$

which, conveniently assembled, give bi-graded algebra homomorphisms.

Lecture 4

Hermitean linear algebra

We discuss Hermitean forms on a complex vector space, the associated symmetric and alternating forms, the induced inner product on the complexified exterior algebra and we introduce the Weil operator.

4.1 The exterior algebra on $V_{\mathbb{C}}^*$

Let things be as in §3.5, i.e. we have, for example,

$$V_{\mathbb{R}}^* \subseteq V_{\mathbb{C}}^* = V'^* \oplus V''^*.$$

Given the natural isomorphism $\Lambda_{\mathbb{R}}(V_{\mathbb{R}}^*) \otimes_{\mathbb{R}} \mathbb{C} \simeq \Lambda_{\mathbb{C}}(V_{\mathbb{C}}^*)$, we have the inclusion of exterior algebras

$$\Lambda_{\mathbb{R}}(V_{\mathbb{R}}^*) \subseteq \Lambda_{\mathbb{C}}(V_{\mathbb{C}}^*)$$

where the complexified one carries the two identical c-conjugation and operational conjugation (3.2) which fix precisely the real exterior algebra.

Given any basis $\{e_j^*\}$ for V'^*, the real vector

$$\boxed{(\tfrac{i}{2})^n \, e_1^* \wedge \overline{e_1^*} \wedge \ldots \wedge e_n^* \wedge \overline{e_n^*} \ \in \ \Lambda_{\mathbb{R}}^{2n}(V_{\mathbb{R}}^*)} \tag{4.1}$$

gives an orientation for $V_{\mathbb{R}}^*$ which is independent of the choice of basis and is therefore considered canonical. See the calculation following Proposition 3.4.1.

Define

$$\boxed{\Lambda^{p,q}(V_{\mathbb{C}}^*) := \Lambda_{\mathbb{C}}^p(V'^*) \otimes_{\mathbb{C}} \Lambda_{\mathbb{C}}^q(V''^*)}$$

which can be viewed as sitting naturally in $\Lambda_{\mathbb{C}}^{p+q}(V_{\mathbb{C}}^*)$.

Its elements can be written as complex linear combinations of vectors of the form

$$f_1 \wedge \ldots \wedge f_p \wedge \overline{g_1} \wedge \ldots \wedge \overline{g_q}, \qquad f_j, \, g_k \in V'^*.$$

We have real linear isomorphisms induced by conjugation

$$\Lambda^{p,q}(V_{\mathbb{C}}^*) \simeq_{\mathbb{R}} \Lambda^{q,p}(V_{\mathbb{C}}^*).$$

We have a natural internal direct sum decomposition

$$\boxed{\Lambda_{\mathbb{C}}^l(V_{\mathbb{C}}^*) = \bigoplus_{p+q=l} \Lambda^{p,q}(V_{\mathbb{C}}^*).}$$

If $V_{\mathbb{R}}^*$ has a metric g^*, then we get the induced metric, still denoted by g^*, and the star operator on $\Lambda_{\mathbb{R}}(V_{\mathbb{R}}^*)$ for the given metric and the orientation (4.1).

4.2 Bases

Let $\{e_1, \ldots, e_n\}$ be a basis for V' and $\{e_1^*, \ldots, e_n^*\}$ be the dual basis for V'^*.

Exercise 4.2.1. Show that

$$\{\overline{e_1}, \ldots, \overline{e_n}\}$$

is a basis for V''.
  Show that

$$\{\overline{e_1^*}, \ldots, \overline{e_n^*}\}$$

is the corresponding dual basis for V''^* where the conjugation is the one associated with the complexification, or equivalently, the operational one.
  Show the analogues of (3.3), (3.4), (3.10) and (3.11) i.e.: show that

$$\boxed{x_j := e_j + \overline{e_j},} \qquad \boxed{y_j := i\,(e_j - \overline{e_j})}$$

form a basis for $V_{\mathbb{R}}$ whose dual basis for $V_{\mathbb{R}}^*$ is given by

$$\boxed{x_j^* = \tfrac{1}{2}\,(e_j^* + \overline{e_j^*}),} \qquad \boxed{y_j^* = \tfrac{1}{2i}\,(e_j^* - \overline{e_j^*})}$$

and that

$$\boxed{e_j = \tfrac{1}{2}(x_j - i\,y_j),} \qquad \boxed{\overline{e_j} = \tfrac{1}{2}(x_j + i\,y_j),}$$

$$\boxed{e_j^* = x_j^* + i\,y_j^*,} \qquad \boxed{\overline{e_j^*} = x_j - i\,y_j^*.}$$

4.3 Hermitean metrics

Definition 4.3.1. (Hermitean forms) A *Hermitean form* on a finite dimensional complex vector space W is a $\mathbb{C}$-bilinear form

$$h : W \times \overline{W} \longrightarrow \mathbb{C}$$

such that

$$h(v, w) = \overline{h(w, v)}, \quad \forall v, w \in W.$$

A *Hermitean metric* on the complex vector space W is a positive definite Hermitean form, i.e. one for which

$$h(v, v) > 0, \quad \forall \, 0 \neq v \in W.$$

Exercise 4.3.2. Verify the following assertions. Let h be a Hermitean form on a complex vector space W. Note that [Weil, 1958], §*I*.2 uses a different convention, where f is anti-linear in the first variable so that there are sign differences in what follows.

The real bilinear form on the real vector space $W = \overline{W}$ given by $S_h = Re \, h : W \times W \longrightarrow \mathbb{R}$ is symmetric.

The form S_h is positive definite iff h is a Hermitean metric.

The real bilinear form $A_h = Im \, h : W \times W \longrightarrow \mathbb{R}$ is anti-symmetric.

We have, dropping the sub-fix "h"

$$S(w, w') = A(i\,w, w') = -A(w, i\,w'), \quad A(w, w') = S(w, i\,w') = -S(i\,w, w'),$$

$$S(i\,w, i\,w') = S(w, w'), \quad A(i\,w, i\,w') = A(w, w').$$

Let S' be a symmetric $\mathbb{R}$-bilinear form on the real vector space W which is invariant under the $\mathbb{C}$-linear automorphism of W given by $w \to i\,w$. Then $A'(w, w') := S'(w, i\,w')$ defines a real alternating form on the real vector space W and the form $S' + i\,A'$ is Hermitean.

Let A'' be an alternating $\mathbb{R}$ form on the real vector space W invariant under $w \to i\,w$. Then $S''(w, w') := A''(i\,w, w')$ defines a symmetric bilinear form and the form $S'' + i\,A''$ is Hermitean.

There are bijections between the following three sets: the set of Hermitean forms on W, the set of real symmetric bilinear forms on W invariant under $w \to i\,w$, the set of real alternating bilinear forms on W invariant under $W \to i\,W$.

Definition 4.3.3. (The alternating bilinear form associated with h)
The *alternating bilinear form* associated with a Hermitean form h is

$$\omega_h := -A_h = -\operatorname{Im} h.$$

A Hermitean form h is a tensor in $W^* \otimes_{\mathbb{C}} \overline{W}^*$.

If $\{e_j\}$ is a basis for W, then we get the dual basis $\{e_j^*\}$ for W^* and the dual basis $\{\epsilon_j^*\}$ for $\overline{W}^*$.

We have that $\epsilon_j^* = \overline{e_j^*}^r$, i.e. $\epsilon_j^*(w) = \overline{e_j^*(w)}$ and we can write

$$h = \sum_{j,k} h_{jk}\, e_j^* \otimes \overline{\epsilon_k^*}^r, \qquad h_{jk} := h(e_j, e_k), \qquad h_{jk} = \overline{h_{kj}}.$$

The slight problem with this set-up is that we wish to perform conjugation operations in order, for example, to give expressions for S_h and ω_h and the expression $\overline{e_j^* \otimes \overline{\epsilon_k^*}}^r = \overline{e_j^*}^r \otimes e_k^*$ does not represent, strictly speaking, an equality in $W^* \otimes_{\mathbb{C}} \overline{W}^*$.

The upshot of Exercise 3.5.6 is that we can, by setting $V_{\mathbb{R}} := W_0$, view W as a V', $\overline{W}$ as a V'' etc. and view h as a $\mathbb{C}$-bilinear map

$$\widetilde{h} : V' \times V'' \longrightarrow \mathbb{C}$$

with, setting $v' := l^{-1}(v)$ (see Exercise 3.5.6) etc.,

$$h(v, w) = \widetilde{h}(v', \overline{w'}).$$

The tensor $\widetilde{h} \in V'^* \otimes V''^*$ can now be viewed in $V_{\mathbb{C}}^* \otimes_{\mathbb{C}} V_{\mathbb{C}}^*$ and as such it can be conjugated, using the operational conjugation, in a way compatible with all the isomorphisms and conjugations considered above.

We are now free to write, with abuse of notation,

$$\boxed{h = \sum_{j,k} h_{jk}\, e_j^* \otimes \overline{e_k^*}^o,} \qquad \boxed{h_{jk} = \overline{h_{kj}}} \qquad (4.2)$$

and we are free to conjugate tensors.

Of course we choose to write $\overline{e_k^*}^o$ simply as $\overline{e_k^*}$.

Exercise 4.3.4. (The Euclidean metric on $\mathbb{C}^n$) Let $\{e_j\}$ be the standard basis for $\mathbb{C}^n$ and define

$$h\left(\sum_j a_j e_j, \sum_k b_k e_k\right) := \sum_j a_j \overline{b}_j.$$

Verify, using the bases of Exercise 4.2.1, that

$$h = \sum_j e_j^* \otimes \overline{e_j^*},$$

$$S_h = Re\,h = \sum_j (x_j^* \otimes x_j^* + y_j^* \otimes y_j^*)$$

$$\omega_h = -Im\,h = -A_h = \sum_j x_j^* \wedge y_j^* = \frac{i}{2}\sum_j e_j^* \wedge \overline{e_j^*}.$$

The ordered basis $\{x_1^*, y_1^*, \ldots, x_n^*, y_n^*\}$ is orthonormal with respect to the Euclidean metric on $\mathbb{C}^{n*} = \mathbb{R}^{2n}$ given by the dual metric S_h^*. Endow this latter space with the orientation (4.1). Verify that

$$\frac{1}{n!}\,\omega_h^n = dV_{S_h^*}.$$

Using the expression (4.2) for the Hermitean form h, we deduce that

$$\boxed{\omega_h = \frac{i}{2}\sum_{j,k} h_{jk}\, e_j^* \wedge \overline{e_k^*},} \qquad \boxed{h_{jk} = \overline{h_{kj}}} \qquad (4.3)$$

The 2-form ω_h is a real $(1,1)$-form. By Exercise 4.3.2, giving a real $(1,1)$-form ω which, as an alternating form on V', is invariant under multiplication by i on V', is equivalent to giving a Hermitean form h_ω. In this case, we have $\omega_{h_\omega} = \omega$.

The Hermitean form is a metric iff $\|h_{jk}\|$ is positive definite (with respect to any basis) which in turn is equivalent to the associated $(1,1)$-form ω_h being *positive*, i.e. being such that $i\,\omega_h(v', \overline{v'}) > 0$ for every $0 \neq v' \in V'$.

The Graham-Schmidt process ensures that if h is a Hermitean metric, then we can find a *unitary basis* for h, i.e. a basis $\{e_j\}$ for V' such that $h(e_j, e_k) = \delta_{jk}$ so that

$$h = \sum_j e_j^* \otimes \overline{e_j^*}$$

and

$$\omega_h = \frac{i}{2}\sum_j e_j^* \wedge \overline{e_j^*} = \sum_j x_j^* \wedge y_j^* \qquad (4.4)$$

from which it is apparent that ω is a real $(1,1)$-form.

We also have

$$S_h = \operatorname{Re} h = \sum_j \left(x_j^* \otimes x_j^* + y_j^* \otimes y_j^* \right)$$

and the orientation (4.1), giving the volume element

$$dV_{S_h^*} = x_1^* \wedge y_1^* \wedge \ldots \wedge x_n^* \wedge y_n^*.$$

A straightforward calculation, i.e. taking the n-th exterior power of the expression (4.3), gives the following

Proposition 4.3.5. *Let h be a Hermitean metric on a complex vector space V'. Using the standard orientation on the space $V_\mathbb{R}^*$ and the metric S_h^* on $V_\mathbb{R}^*$, we have the equality:*

$$\frac{1}{n!}\,\omega^n = dV_{S_h^*}.$$

Remark 4.3.6. (Wirtinger's Inequality) If h is a Hermitean metric on a complex vector space W, then we get two positive top forms on the real vector space underlying W^*, i.e. $\frac{1}{n!}\omega_h^n$ and $dV_{S_h^*}$. By Proposition 4.3.5 they coincide.

If $U \subseteq W$ is a real vector subspace of dimension $2k$, then we have two positive forms on U : the restriction $S_{h|U}^*$ and the restriction of $\frac{1}{k!}\omega_{h|U}^k$.

Wirtinger's Inequality states that

$$\frac{1}{k!}\,\omega_{h|U}^k \leq S_{h|U}^*$$

and equality holds iff $U \subseteq W$ is a complex subspace. See [Mumford, 1970], p. 88 and the discussion that follows, which culminates with [Mumford, 1970], Theorem 5.35, concerning volume-minimizing submanifolds of $\mathbb{P}^n$.

4.4   The inner product and the $\star$ operator on the complex-ified exterior algebra $\Lambda_\mathbb{C}(V_\mathbb{C}^*)$

Let h be a Hermitean metric on V'. The metric S_h^* on $V_\mathbb{R}^*$ induces a metric, g^*, on $\Lambda_\mathbb{R}(V_\mathbb{R}^*)$ and, using the orientation (4.1), the $\star$ operator on $\Lambda_\mathbb{R}(V_\mathbb{R}^*)$.

The metric g^* is a symmetric, positive definite bilinear form on $\Lambda_{\mathbb{R}}(V_{\mathbb{R}}^*)$. We consider its complexification, i.e. the $\mathbb{C}$-bilinear form $g^* \otimes Id_{\mathbb{C}}$ on $\Lambda_{\mathbb{C}}(V_{\mathbb{C}}^*)$, and define a Hermitean metric $\langle\,,\,\rangle$ on $\Lambda_{\mathbb{C}}(V_{\mathbb{C}}^*)$ by setting

$$\boxed{\langle u, v \rangle := (g^* \otimes Id_{\mathbb{C}})\,(u, \overline{v}).} \qquad (4.5)$$

Exercise 4.4.1. Verify that $\langle\,,\,\rangle$ is a Hermitean metric, that the spaces $\Lambda_{\mathbb{C}}^l(V_{\mathbb{C}}^*)$ are mutually orthogonal, that the spaces $\Lambda^{p,q}(V_{\mathbb{C}}^*)$ are mutually orthogonal and that, given a h-unitary basis for V':

$$\| e_J^* \wedge \overline{e_K^*} \| = 2^{|J|+|K|}. \qquad (4.6)$$

(See also Exercise 1.1.1.)

Definition 4.4.2. The $\star$ *operator* on $\Lambda_{\mathbb{C}}(V_{\mathbb{C}}^*)$ is defined to be the $\mathbb{C}$-linear extension of the $\star$ operator on $\Lambda_{\mathbb{R}}(V_{\mathbb{R}}^*)$.

Exercise 4.4.3. Show that

$$u \wedge \overline{\star v} = \langle u, v \rangle\, dV_{S_h^*}, \qquad \forall\, u,\, v \in \Lambda_{\mathbb{C}}^l(V_{\mathbb{C}}^*). \qquad (4.7)$$

Using (4.7), show that this new, extended $\star$ operator gives isometries

$$\Lambda^{p,q}(V_{\mathbb{C}}^*) \simeq_{\mathbb{C}} \Lambda^{n-q,n-p}(V_{\mathbb{C}}^*). \qquad (4.8)$$

In particular,

$$\star\, \pi^{p,q} = \pi^{n-q,n-p}\, \star. \qquad (4.9)$$

Finally, observe that since the real dimension of $V_{\mathbb{R}}$ is even, (1.4) implies that

$$\star\star|_{\Lambda^{p,q}(V_{\mathbb{C}}^*)} = (-1)^{p+q} Id_{\Lambda^{p,q}(V_{\mathbb{C}}^*)}. \qquad (4.10)$$

The explicit form of this new extended $\star$ operator using the real S_h^*-orthonormal ordered basis $\{x_1^*, y_1^*, \ldots, x_n^*, y_n^*\}$ associated with a h-unitary basis e_j is identical to the non-complexified one.

The expression of the $\star$ operator using the e_j^*, $\overline{e_j^*}$ basis can be found in [Weil, 1958], pp. 19-20.

4.5   The Weil operator

The Weil operator can be defined for any Hodge structure. It is very convenient in view of the definition and the use of polarizations of pure Hodge structures; see §7.1.

Here we look at the exterior algebra $\Lambda_{\mathbb{C}}(V_{\mathbb{C}}^*)$ which is a direct sum of the weight l pure Hodge structures $\Lambda_{\mathbb{C}}^l(V_{\mathbb{C}}^*)$; see §4.1.

Definition 4.5.1. (The Weil operator) The *Weil operator* is the complex linear isomorphism $\Lambda_{\mathbb{C}}(V_{\mathbb{C}}^*) \longrightarrow \Lambda_{\mathbb{C}}(V_{\mathbb{C}}^*)$ induced by multiplication by i on V', i.e.

$$C := \sum_{p,q} i^{p-q} \, \pi^{p,q}.$$

The Weil operator is real, i.e. it preserves the real subspace $\Lambda_{\mathbb{R}}(V_{\mathbb{R}})$.

By (4.9), we have

$$C \star - \star C = 0. \tag{4.11}$$

Since $\dim_{\mathbb{R}} V_{\mathbb{R}}^*$ is even,

$$\star\star = \sum_{l=0}^{2n} (-1)^l \pi^l = w$$

where w is the so-called *de Rham operator* w.

The following relations follow:

$$w = \star\star = C^2,$$

$$\star^{-1} = w\star = \star w, \qquad C^{-1} = wC = Cw.$$

Lecture 5

The Hodge theory of Hermitean manifolds

We discuss Hermitean metrics on complex manifolds, the Δ' and Δ'' Laplacians, the Δ' and Δ'' harmonic forms, the corresponding Hodge Theory on a compact complex manifold, including Kodaira-Serre Duality.

5.1 Hermitean metrics on complex manifolds

A Hermitean metric h on a complex manifold X is the assignment of a Hermitean metric

$$h(-,-)_x : T'_{X,x} \otimes_{\mathbb{C}} \overline{T'}_{X,x} \longrightarrow \mathbb{C} \tag{5.1}$$

for every $x \in X$ varying smoothly with x, i.e. such that the functions

$$h_{jk}(z) := (\partial_{z_j}, \partial_{z_k})$$

are smooth on the open set U.

Recalling the discussion culminating with (4.2), using the local chart $(U; z)$, the Hermitean metric h can be expressed on U in tensor form as

$$h = \sum_{jk} h_{jk}(z) dz_j \otimes d\bar{z}_k. \tag{5.2}$$

The real $(1,1)$-form

$$\omega = \omega_h = -Im\, h \in A^{1,1}(X)$$

is called the *associated $(1,1)$-form* of the metric h.

Using the chart U, it can be written as

$$\omega = \frac{i}{2} \sum_{jk} h_{jk}(z)\, dz_j \wedge d\bar{z}_k, \qquad h_{jk} = \overline{h_{kj}}.$$

The metric h can be recovered from the associated $(1,1)$ form.

Exercise 5.1.1. (**Existence of Hermitean metrics on complex manifolds**) Show that any complex manifold admits Hermitean metrics on it.

Exercise 5.1.2. (**Restriction of Hermitean metrics**) Let $f : X^n \longrightarrow Y^m$ a holomorphic map of complex manifolds of the indicated dimensions $n \leq m$ such that $df : T'_X \longrightarrow f^* T'_Y$ is everywhere of rank n, i.e. injective. Let h_Y be a Hermitean metric on Y. Show that one can induce a Hermitean metric h_X on X such that $\omega_{h_X} = f^* \omega_{h_Y}$. In particular, if $f : X \longrightarrow Y$ is an embedding of complex manifolds, then $h_X := h_{Y|X}$ is a Hermitean metric and $\omega_{h_X} = \omega_{h_Y|X}$. See [Griffiths and Harris, 1978], p. 29.

An important related fact is *Wirtinger's Theorem*, [Griffiths and Harris, 1978], p. 31, [Mumford, 1970], p. 88. See also Proposition 4.3.5 and Remark 4.3.6.

Exercise 5.1.3. (**Wirtinger's Theorem**) Prove Wirtinger's Theorem: let $Y \subseteq X$ be a complex submanifold of complex dimension k, h be a Hermitean metric on X and ω be the associated form; then

$$vol(Y) = \frac{1}{k!} \int_Y \omega_{|Y}^k. \qquad (5.3)$$

Note the following special feature of complex geometry expressed by (5.3): the volume of Y is expressed as the integral over Y of a globally defined differential form on X. This does not occur in general in the real case. See [Griffiths and Harris, 1978], p. 31.

Exercise 5.1.4. (**The Fubini-Study metric on $\mathbb{P}^n$**) Verify all the following assertions. See [Griffiths and Harris, 1978], p. 30. See also [Mumford, 1970], pp. 86-87.

Let $[x_0 : \ldots : x_n]$ be homogeneous coordinates on $\mathbb{P}^n$. The expression $log(|x_0|^2 + \ldots + |x_n|^2)$ is well-defined on $\mathbb{C}^{n+1} \setminus \{0\}$. The differential form

$$\omega' := \frac{i}{2\pi} d' d'' \, log(|x_0|^2 + \ldots + |x_n|^2) \in A^{1,1}(\mathbb{C}^{n+1} \setminus \{0\})$$

is $\mathbb{C}^*$-invariant. Show that it descends to a $(1,1)$-form $\omega \in A^{1,1}(\mathbb{P}^n)$. On the chart U^0 (and on any chart, keeping track of indices) we have

$$\omega = \frac{i}{2\pi} d' d'' \log(1 + |z_1|^2 + \ldots + |z_n|^2)$$

$$= \frac{i}{2\pi} \left[\frac{\sum_j dz_j \wedge d\bar{z}_j}{1 + \sum_j |z_j|^2} - \frac{\left(\sum_j \bar{z}_j dz_j\right) \wedge \left(\sum_j z_j d\bar{z}_j\right)}{\left(1 + \sum_j |z_j|^2\right)^2} \right].$$

At the point $[1 : 0 : \ldots : 0]$

$$\omega = \frac{i}{2\pi} \sum_j dz_j \wedge d\bar{z}_j > 0$$

so that ω is the associated $(1,1)$-form for a Hermitean metric defined on $\mathbb{P}^n$. This metric is called the *Fubini-Study* metric of $\mathbb{P}^n$.

Show that $d\omega = 0$, i.e. the Fubini Study metric is Kähler.

Show that $\int_{\mathbb{P}^n} \omega^n = 1$.

Show that $[\omega] \in H^2(X, \mathbb{R})$ is Poincaré dual to the homology class $\{H\} \in H_{2n-2}(X, \mathbb{R})$ associated with a hyperplane $H \subseteq \mathbb{P}^n$ and also that ω is the curvature form associated with a Hermitean metric on the hyperplane bundle L_1 which is therefore *positive* in the complex differential geometric sense and *ample* in the algebraic geometric sense. Finally, show that $[\omega]$ is the first Chern class with $\mathbb{R}$ coefficients of L_1 (this requires some non-trivial unwinding of the definitions; see [Griffiths and Harris, 1978], p. 141).

5.2 The Hodge theory of a compact Hermitean manifold

Let (X, h), be a Hermitean manifold, i.e. a complex manifold X endowed with a Hermitean metric h.

The smooth manifold underlying X carries the natural orientation given by the complex structure on X. See §3.4.

The metric h gives rise to a Hermitean metric $\langle \, , \, \rangle$ on the exterior algebra bundle $\Lambda(T_X^*(\mathbb{C}))$. See (4.5).

We extend, as in Definition 4.4.2, the $\star$ operator to complex-valued forms and we have

$$u \wedge \overline{\star v} = \langle u, v \rangle \, dV, \tag{5.4}$$

where dV is the Hermitean volume element, i.e. the one associated with the orientation and the Riemannian metric associated with the Hermitean metric. See Exercise 4.4.3.

One has that

$$\star : \Lambda_{\mathbb{C}}^{p,q}(T_X^*) \longrightarrow \Lambda_{\mathbb{C}}^{n-q,n-p}(T_X^*)$$

is a complex linear isometry.

We will consider formal adjoints with respect to the following metric.

Definition 5.2.1. Whenever the integral converges, e.g. for compactly supported forms, define

$$\boxed{\langle\langle u, v\rangle\rangle = \int_X \langle u, v\rangle \, dV} = \int_X u \wedge \overline{\star v}.$$

Definition 5.2.2. (The d' and d'' Laplacians)

$$\boxed{d'^{\star} = -\star d'' \star,} \qquad \boxed{d''^{\star} = (d'')^{\star} = -\star d' \star}$$

$$\boxed{\Delta' = d'd'^{\star} + d'^{\star}d',} \qquad \boxed{\Delta'' = d''d''^{\star} + d''^{\star}d''.}$$

The definition is motivated by the following

Exercise 5.2.3. Show, as in Proposition 2.1.5, that $d'^{\star}$ is the formal adjoint to d' and that $d''^{\star}$ is the formal adjoint to d''. Show that Δ' and Δ'' are self-adjoint.

Definition 5.2.4. (Harmonic (p,q)-forms) The space of Δ'-*harmonic* (p,q)-*forms* of X is

$$\boxed{\mathcal{H}_{\Delta'}^{p,q}(X) := \mathrm{Ker}\,(\,\Delta' : A^{p,q}(X) \longrightarrow A^{p,q}(X)\,)}$$

and the space of Δ''-*harmonic* (p,q)-*forms* of X is

$$\boxed{\mathcal{H}_{\Delta''}^{p,q}(X) := \mathrm{Ker}\,(\Delta'' : A^{p,q}(X) \longrightarrow A^{p,q}(X)\,).}$$

These spaces depend on h.

The following results are the complex analytic versions of the Hodge Orthogonal Decomposition Theorem, of the Hodge Isomorphism Theorem and of the Poincaré Duality Theorem. One replaces the de Rham Cohomology with the Doulbeault cohomology and Δ-harmonic forms with Δ''-harmonic forms etc.

Theorem 5.2.5. *(Hodge Theory of compact complex manifolds)* Let (X, h) be a <u>compact</u> Hermitean manifold. Then for every bi-degree (p, q) :

(a) (Hodge Orthogonal Decompositions for d' and d'') there are orthogonal direct sum decompositions

$$\boxed{A^{p,q}(X) = \mathcal{H}^{p,q}_{\Delta'}(X) \overset{\perp}{\oplus} d'(A^{p-1,q}(X)) \overset{\perp}{\oplus} d'^\star(A^{p+1,q}(X)),}$$

$$\boxed{A^{p,q}(X) = \mathcal{H}^{p,q}_{\Delta''}(X) \overset{\perp}{\oplus} d''(A^{p,q-1}(X)) \overset{\perp}{\oplus} d''^\star(A^{p,q+1}(X)).}$$

(b) (Hodge Isomorphisms for d' and d'') there are isomorphisms of <u>finite</u> dimensional complex vector spaces

$$\boxed{\mathcal{H}^{p,q}_{\Delta'}(X) \simeq H^{p,q}_{d'}(X),}$$

$$\boxed{\mathcal{H}^{p,q}_{\Delta''}(X) \simeq H^{p,q}_{d''}(X).}$$

(c) (Kodaira-Serre Duality) the complex bilinear pairings

$$H^{p,q}_{d'}(X) \times H^{n-p,n-q}_{d'}(X) \longrightarrow \mathbb{C}, \qquad (u, v) \longrightarrow \int_X u \wedge v,$$

$$H^{p,q}_{d''}(X) \times H^{n-p,n-q}_{d''}(X) \longrightarrow \mathbb{C}, \qquad (u, v) \longrightarrow \int_X u \wedge v,$$

are non-degenerate dualities. In particular, there is a canonical isomorphism

$$\boxed{H^{n-q}(X, \Omega^{n-p}_X) \simeq H^q(X, \Omega^p_X)^\vee.}$$

Exercise 5.2.6. Prove (b) and (c) using (a) as in Theorems 2.3.7 and 2.3.8.

Lecture 6

Kähler manifolds

We discuss the Kähler condition, the fundamental identities of Kähler geometry, the Hodge Decomposition via the $d'd''$-Lemma and Bott-Chern cohomology (as in [Demailly, 1996]) and some topological implications of the Hodge Decomposition.

It is on compact Kähler manifolds that Hodge theory becomes a formidable tool which highlights some of the amazing properties that these manifolds enjoy: non-vanishing of even Betti numbers, parity of the odd Betti numbers, Kodaira-Serre symmetry and Hodge symmetry for the Hodge numbers $h^{p,q}$. See §6.4.

It is costumary to denote a Hermitean manifold (X, h) also by (X, ω), where $\omega = \omega_h$.

6.1 The Kähler condition

Definition 6.1.1. (Kähler metric/manifold) A Hermitean metric h on a complex manifold X is called *Kähler* if $d\omega = 0$.

A complex manifold X is said to be *Kähler* if it admits a Kähler metric.

Definition 6.1.2. (Projective manifolds) A complex manifold is said to be *projective* if it admits a closed holomorphic embedding in some projective space.

Example 6.1.3. Any Riemann surface is automatically Kähler. Any compact Riemann surface is projective.

By Exercise 5.1.4, $\mathbb{P}^n$ is Kähler. By Exercise 5.1.2, any projective manifold is Kähler.

Any compact complex torus $\mathbb{C}^n/\Lambda$, $\mathbb{Z}^{2n} \simeq \Lambda \subseteq \mathbb{C}^n$ a full lattice, is Kähler. However, "most" tori of complex dimension at least two are not projective due to the *Riemann conditions*. See [Griffiths and Harris, 1978], pp. 300-307.

The two examples that follow, the Hopf surface and the Iwasawa threefold, are compact complex manifolds which are not Kähler in view of some of the special properties of compact Kähler manifolds that we will establish later in this lecture.

The Hopf surface [Demailly, 1996], §5.7 is a compact complex surface which is not Kähler, in fact its first Betti number is one, i.e. it is odd, and this is not possible on a compact Kähler manifold; see Theorem 6.4.2.

The Iwasawa manifold [Demailly, 1996], §8.10 is a compact complex threefold which is not Kähler since it admits holomorphic 1-forms which are not d-closed, an impossibility on a compact Kähler manifold; see Theorem 6.4.1.

Let (X, ω) be a Hermitean manifold. Since ω is a real $(1,1)$-form, the three conditions

$$d\omega = 0, \qquad d'\omega = 0, \qquad d''\omega = 0,$$

are equivalent to each other and to the condition

$$\frac{\partial h_{jk}}{\partial z_l} = \frac{\partial h_{lk}}{\partial z_j}, \quad 1 \le j, k, l \le n.$$

Let (X, ω) be a Kähler manifold. By Proposition 3.5.6

$$dV = \frac{1}{n!}\,\omega^n,$$

where dV is the volume element associated with the Riemannian metric associated with h and the canonical orientation.

There is an important difference from the Hermitean case: the right-hand side being closed defines a cohomology class which cannot be exact on a compact manifold because of Stokes' Theorem and the fact that the integral of the left-hand-side cannot be zero. This implies that all the relevant powers of ω define non-zero cohomology classes. We have the following remarkable consequences for the topology of a compact Kähler manifold.

Theorem 6.1.4. *Let X be a compact Kähler manifold. Then, for every $0 \leq k \leq n$,*

$$b_{2k}(X) = \dim_{\mathbb{R}} H^{2k}(X, \mathbb{R}) > 0.$$

The fundamental class $[V] \in H^{2n-2k}$ of a closed analytic subvariety of X of dimension $\dim_{\mathbb{C}}(V) = k$ is non-zero.

Proof. Let h be a Kähler metric and ω the associated $(1,1)$-form. Since ω is d-closed, so are ω^k, for every $k \geq 0$. Since $\frac{1}{n!} \int_X \omega^n = vol(X) \neq 0$, Stokes' Theorem implies that the cohomology class $[\omega^n]$ is non-zero and so are all the $[\omega^k]$, $0 \leq k \leq n$, due to the obvious relation $[\omega^n] = [\omega^k] \cup [\omega^{n-k}]$.

For the second statement, argue as follows. By Exercise 5.1.3, and by the properties of the current of integration along V $\int_V$ (see Remark 1.4.15), we have that the pairing

$$\left([\int_V], [\omega^k] \right) = \int_{V_{reg}} \omega^k_{|V_{reg}} > 0,$$

so that $[V] = [\int_V] \neq 0$. $\qquad\square$

Let (X, h) be a Hermitean manifold. Let g be the associated Riemannian metric on the underlying smooth manifold, ∇ be the Levi-Civita connection on the real tangent bundle of X associated with g. In fact, every tensor bundle inherits such a connection.

Let $J : T_X(\mathbb{R}) \longrightarrow T_X(\mathbb{R})$ be the quasi complex structure on X, i.e. the associated automorphism of the real tangent bundle. Using a holomorphic local chart $(U; z)$ around $x \in X$, we have $J(\partial_{x_j}) = \partial_{y_j}$ and $J(\partial_{y_j}) = -\partial_{x_j}$.

Exercise 6.1.5. Show that assigning a Hermitean metric h on the complex manifold X is equivalent to assigning a J-invariant Riemannian metric on the underlying smooth manifold.

Theorem 6.1.6. (Characterization of Kähler metrics) *Let (X, h) be a Hermitean variety. The following conditions are equivalent.*

(a) $d\omega = 0$, i.e. h is Kähler;

(b) the types of the complexified tangent vectors are preserved under parallel transport;

(c) for every real parallel vector field η along a smooth curve γ, $J\eta$ is parallel along γ;

(d) $\nabla J = 0$, i.e. the almost complex structure is parallel;

(e) $\nabla \omega = 0$, i.e. the associated $(1,1)$-form is parallel;

(f) h admits a potential locally on X, i.e. a function φ such that, in local holomorphic coordinates, $h_{jk} = \frac{\partial^2 \varphi}{\partial z_j \partial \bar{z}_k}$;

(g) for every $x \in X$ there is a holomorphic chart $(U; z)$ centered at x such that $h_{jk}(z) = \delta_{jk} + O(|z|^2)$;

(h) the torsion of the Hermitean metric h is zero.

Proof. See [Mok, 1989], p. 18 and [Griffiths and Harris, 1978], p. 107. See also [Demailly, 1996], §5.8. See also Exercise 6.1.7 below. $\qquad\square$

Exercise 6.1.7. Show that (g) implies (a) by directly differentiating the expression

$$\omega = \frac{i}{2} \sum_{j,k} (\delta_{jk} + [2]) \, dz_j \wedge d\bar{z}_k$$

for ω at a given point.

Show that (a) implies (g) as follows.

Show that the Graham-Schmidt process implies the existence of holomorphic coordinates at any given point x such that $h_{jk}(x) = \delta_{jk}$ so that

$$\omega = \frac{i}{2} \sum_{j,k,l} (\delta_{jk} + a_{jkl} z_k + a_{jk\bar{l}} \bar{z}_k) \, dz_j \wedge d\bar{z}_k = \frac{i}{2} \sum_{j,k} h_{jk}(z) \, dz_j \wedge d\bar{z}_k.$$

Show, using $h_{jk}(x) = \overline{h_{kj}(x)}$, that

$$\overline{a_{jkl}} = a_{kjl}.$$

Show, using $d\omega = 0$, that

$$a_{jkl} = a_{lkj}.$$

Use these relations to show that, by setting

$$b_{juv} = -a_{vju},$$

the change of coordinates defined by the equality

$$z_j = w_j + \frac{1}{2} \sum_{u,v} b_{juv} \, w_u w_v$$

gives

$$\omega = \frac{i}{2} \sum_{j,k} (\delta_{jk} + [2])\, dw_j \wedge d\overline{w}_k.$$

Remark 6.1.8. The characterization $(a) = (g)$ is used in §6.2 to prove the fundamental identities of Kähler geometry by first showing them on $\mathbb{C}^n$ with the Euclidean metric and then by observing that these identities involve only the metric and its first derivatives so that proving the Euclidean case is enough. See [Griffiths and Harris, 1978], p. 115.

6.2 The fundamental identities of Kähler geometry

Definition 6.2.1. (**The operator L and its adjoint**). Let (X,h) be a Hermitean manifold and define

$$L : A^\bullet(X) \longrightarrow A^{\bullet+2}(X), \quad L(u) := \omega \wedge u,$$
$$L^\star : A^\bullet(X) \longrightarrow A^{\bullet-2}(X), \quad L^\star(u) := \star^{-1} L \star$$

The operators d', d'', $d'^\star$, $d''^\star$ Δ', Δ'', L and $L^\star$ have bi-degree $(1,0)$, $(0,1)$, $(-1,0)$, $(0,-1)$, $(0,0)$, $(0,0)$, $(1,1)$ and $(-1,-1)$.

The *commutator* $[A,B]$ of two operators of bi-degree (a',a'') and (b',b'') and of total degree $a = a' + a''$, $b = b' + b''$ is defined as

$$[A,B] := AB - (-1)^{ab} BA. \tag{6.1}$$

Theorem 6.2.2. *(Fundamental identities of Kähler geometry)*
Let (X,h) be a Kähler manifold. Then

$$\boxed{[d''^\star, L] = i\, d',} \quad \boxed{[d'^\star, L] = -i\, d'',} \quad \boxed{[L^\star, d''] = -i\, d'^\star,} \quad \boxed{[L^\star, d'] = i\, d''^\star.}$$
$$\tag{6.2}$$

$$\boxed{[d', d''^\star] = 0,} \quad \boxed{[d'', d'^\star] = 0,} \tag{6.3}$$

$$\boxed{\Delta = 2\Delta' = 2\Delta''}. \tag{6.4}$$

In addition, Δ, Δ' and Δ'' commute with d, $\star$, d', d'', $d'^\star$, $d''^\star$, L and $L^\star$.

Finally, Δ preserves the (p,q)-decomposition, i.e. Δ commutes with $\pi^{p,q}$.

Proof. The first and second relations (6.2) are conjugate to each other and so are the third and fourth.

The first and third are adjoint to each other.

It follows that the four relations are equivalent to each other.

We prove the fourth relation in the guided Exercise 6.2.3.

We now prove (6.3). Since the two relations are conjugate to each other it is enough to show the first one. We have

$$i\,(d'd''^{\star} + d''^{\star}d') = d'(i\,d''^{\star}) + (i\,d''^{\star})d' =$$

$$d'(L^{\star}d' - d'L^{\star}) + (L^{\star}d' - d'L^{\star})d' = d'L^{\star}d' - d'^2L^{\star} + L^{\star}d'^2 - d'L^{\star}d' = 0.$$

We now prove (6.4). Firstly, we have the equality:
$$\begin{aligned}\Delta_d &= (d' + d'')(d'^{\star} + d''^{\star}) + (d'^{\star} + d''^{\star})(d' + d'')\\ &= (d'd'^{\star} + d'^{\star}d') + (d''d''^{\star} + d''^{\star}d'') + (d'd''^{\star} + d''^{\star}d') + (d''d'^{\star} + d'^{\star}d'')\\ &= \Delta_{d'} + \Delta_{d''}.\end{aligned}$$

Secondly, we prove that $\Delta_{d'} = \Delta_{d''}$ as follows:
$$\begin{aligned}-i\,\Delta_{d'} &= -i\,d'd'^{\star} - i\,d'^{\star}d' = d'(-i\,d'^{\star}) + (-i\,d'^{\star})d'\\ &= d'(L^{\star}d'' - d''L^{\star}) + (L^{\star}d'' - d''L^{\star})d'\\ &= d'L^{\star}d'' - d'd''L^{\star} + L^{\star}d''d' - d''L^{\star}d'\\ &= d'L^{\star}d'' + d''d'L^{\star} - L^{\star}d'd'' - d''L^{\star}d'\\ &= -(d''L^{\star}d' - d''d'L^{\star} + L^{\star}d'd'' - d'L^{\star}d'')\\ &= -[d''(L^{\star}d' - d'L^{\star}) + (L^{\star}d' - d'L^{\star})d'']\\ &= -[d''(i\,d''^{\star}) + (i\,d''^{\star})d''] = -i\,\Delta_{d''}.\end{aligned}$$

(6.3) implies that Δ' commutes with d'' and that Δ'' commutes with d'.

(6.4) implies that Δ commutes with d', d'' and hence with $d = d' + d''$.

Since Δ is self-adjoint and commutes with d, d' and d'', Δ commutes with $d^{\star}$, $d'^{\star}$ and $d''^{\star}$.

One verifies that Δ commutes with L and $L^{\star}$ by using (6.2).

Finally, since Δ' (and Δ'') are of bi-degree $(0, 0)$, (6.4) implies that Δ is also of type $(0, 0)$, i.e. that it preserves bi-degrees. $\square$

Exercise 6.2.3. (**Proof of the commutation relations**) Prove that

$$[L^{\star}, d'] = i\,d''^{\star} \tag{6.5}$$

as follows. See [Griffiths and Harris, 1978], pp. 111-114.

By Remark 6.1.8 it is enough to prove the relation for the Euclidean metric on $\mathbb{C}^n$.

Define operators

$$\wedge_j \; : \; A_c^{p,q} \longrightarrow A_c^{p+1,q}, \qquad \wedge_j(u) := dz_j \wedge u,$$

$$\overline{\wedge}_j \; : \; A_c^{p,q} \longrightarrow A_c^{p,q+1}, \qquad \overline{\wedge}_j(u) := d\bar{z}_j \wedge u,$$

$$\partial_j \; : \; A_c^{p,q} \longrightarrow A_c^{p,q}, \qquad \partial_j(u_{JK} \, dz_J \wedge d\bar{z}_K) := \frac{\partial u_{JK}}{\partial z_j} dz_J \wedge d\bar{z}_K$$

$$\overline{\partial}_j \; : \; A_c^{p,q} \longrightarrow A_c^{p,q}, \qquad \overline{\partial}_j(u_{JK} \, dz_J \wedge d\bar{z}_K) := \frac{\partial u_{JK}}{\partial \bar{z}_j} dz_J \wedge d\bar{z}_K.$$

We have

$$L = \frac{i}{2} \sum_j \wedge_j \circ \overline{\wedge}_j$$

$$d' = \sum_j \partial_j \circ \wedge_j, \qquad d'' = \sum_j \overline{\partial}_j \circ \overline{\wedge}_j.$$

Note that, concerning adjoints, we have

$$(A \circ B)^\star = B^\star \circ A^\star, \qquad (\lambda A)^\star = \overline{\lambda} A^\star.$$

Each of these operators has its own adjoint:

$$\wedge_j^\star, \quad \overline{\wedge}_j^\star, \quad \partial_j^\star, \quad \overline{\partial}_j^\star,$$

$$L^\star = -\frac{i}{2} \sum_j \overline{\wedge}_j^\star \circ \wedge_j^\star,$$

$$d'^\star = \sum_j \wedge_j^\star \circ \partial_j^\star, \qquad d''^\star = \sum_j \overline{\wedge}_j^\star \circ \overline{\partial}_j^\star.$$

We may omit the "∘."

Use the definition of the inner product via integration and integration by parts to show that

$$\partial_j^\star = -\overline{\partial}_j, \qquad \overline{\partial}_j^\star = -\partial_j. \tag{6.6}$$

Note, also using (6.6), that

∂_j and $\overline{\partial}_k$ commute with each other, with $\wedge_j$ and $\overline{\wedge}_k$, with $\wedge_j^\star$ and $\overline{\wedge}_k^\star$.

$$\tag{6.7}$$

The goal of what follows is to show that (6.12), (6.13) and (6.17) below hold.

In what follows f and g are compactly supported smooth functions on $\mathbb{C}^n$.

Use the orthogonality properties of the inner product on the exterior algebra (see §1.1) to verify that for every index $j \notin J$:

$$\langle\!\langle \wedge_j^\star(f\, dz_J \wedge d\overline{z}_K)\, , g\, dz_L \wedge d\overline{z}_M \rangle\!\rangle = \langle\!\langle f\, dz_J \wedge d\overline{z}_K)\, , g\, dz_j \wedge dz_L \wedge d\overline{z}_M \rangle\!\rangle = 0$$

for every pair of multi-indices L and M and deduce that

$$\wedge_j^\star(f\, dz_J \wedge d\overline{z}_K) = 0, \qquad j \notin J, \tag{6.8}$$

and, in a similar way, that

$$\overline{\wedge}_j^\star(f\, dz_J \wedge d\overline{z}_K) = 0, \qquad j \notin K. \tag{6.9}$$

Use the fact that $||dz_j||^2 = 2$ at every point of $\mathbb{C}^n$ (see Exercise 4.4.1) and the orthogonality properties of the inner product on the exterior algebra to show that

$$\langle\!\langle \wedge_j^\star(f\, dz_j \wedge dz_J \wedge d\overline{z}_K)\, , g\, dz_L \wedge d\overline{z}_M \rangle\!\rangle$$
$$= \langle\!\langle f\, dz_j \wedge dz_J \wedge d\overline{z}_K)\, , g\, dz_j \wedge dz_L \wedge d\overline{z}_M \rangle\!\rangle$$
$$= 2\, \langle\!\langle f\, dz_j \wedge dz_J \wedge d\overline{z}_K\, , g\, dz_L \wedge d\overline{z}_M \rangle\!\rangle$$

for every pair of multi-indices L and M.

Deduce that

$$\wedge_j^\star(f\, dz_j \wedge dz_J \wedge d\overline{z}_K) = 2\, f\, dz_J \wedge d\overline{z}_K, \qquad j \notin J \tag{6.10}$$

and, in a similar way, that

$$\overline{\wedge}_j^\star(f\, d\overline{z}_j \wedge dz_J \wedge d\overline{z}_K) = 2\, f\, dz_J \wedge d\overline{z}_K, \qquad j \notin K. \tag{6.11}$$

Clearly, the left-hand sides of (6.10) and (6.11) are zero if, respectively, $j \in J, j \in K$.

Note that (6.9) and (6.11) imply

$$\wedge_j \overline{\wedge}_k^\star + \overline{\wedge}_k^\star \wedge_j = 0 \tag{6.12}$$

and, by conjugation,

$$\overline{\wedge}_j \wedge_k^\star + \wedge_k^\star \overline{\wedge}_j = 0. \tag{6.13}$$

Use (6.8) and (6.10) to verify that

$$\wedge_j^\star \wedge_j (f \, dz_J \wedge d\bar{z}_K) = \begin{cases} 0 & \text{if } j \in J \\ 2f \, dz_J \wedge d\bar{z}_K & \text{if } j \notin J \end{cases}$$

and that

$$\wedge_j \wedge_j^\star (f \, dz_J \wedge d\bar{z}_K) = \begin{cases} 2f \, dz_J \wedge d\bar{z}_K & \text{if } j \in J \\ 0 & \text{if } j \notin J. \end{cases}$$

Deduce that

$$\wedge_j \wedge_j^\star + \wedge_j^\star \wedge_j = 2 \, Id. \qquad (6.14)$$

Let $j \neq k$. Use (6.10) to show that

$$\wedge_j^\star \wedge_k (f \, dz_j \wedge dz_J \wedge d\bar{z}_K) = - \wedge_k \wedge_j^\star (f \, dz_j \wedge dz_J \wedge d\bar{z}_K). \qquad (6.15)$$

Let $j \neq k$. Show that

$$\wedge_j^\star \wedge_k (dz_J \wedge d\bar{z}_K) = \wedge_k \wedge_j^\star (dz_J \wedge d\bar{z}_K) = 0, \qquad j \notin J. \qquad (6.16)$$

Note that (6.14), (6.15) and (6.16) imply

$$\wedge_j^\star | \wedge_k + \wedge_k \wedge_j^\star = 2 \, \delta_{jk} \, Id. \qquad (6.17)$$

Verify (6.5) using (6.6), (6.7), (6.12) and (6.17) as follows.

$$L^\star d' = -\frac{i}{2} \sum_{j,k} \overline{\wedge}_j^\star \wedge_j^\star \partial_k \wedge_k = \text{ by (6.7) } = -\frac{i}{2} \sum_{j,k} \partial_k \overline{\wedge}_j^\star \wedge_j^\star \wedge_k$$

$$= -\frac{i}{2} \left[\left(\sum_{j=k} \partial_j \overline{\wedge}_j^\star \wedge_j^\star \wedge_j \right) + \left(\sum_{j \neq k} \partial_k \overline{\wedge}_j^\star \wedge_j^\star \wedge_k \right) \right] = \text{ by (6.17)}$$

$$= -\frac{i}{2} \left[\left(\sum_j -\partial_j \overline{\wedge}_j^\star \wedge_j \wedge_j^\star + 2 \sum_j \partial_j \overline{\wedge}_j^\star \right) - \left(\sum_{j \neq k} \partial_k \overline{\wedge}_j^\star \wedge_k \wedge_j^\star \right) \right]$$

$$= \text{ by (6.12) } = -\frac{i}{2} \sum_j \partial_j \wedge_j \overline{\wedge}_j^\star \wedge_j^\star - i \sum_{j=k} \partial_j \overline{\wedge}_j^\star - \frac{i}{2} \sum_{j \neq k} \partial_k \wedge_k \overline{\wedge}_j^\star \wedge_j^\star$$

$$= -\frac{i}{2} \sum_{j,k} \partial_k \wedge_k \overline{\wedge}_j^\star \wedge_j^\star - i \sum_j \partial_j \overline{\wedge}_j^\star = d' L^\star + i \sum_j (-\partial_j) \overline{\wedge}_j^\star$$

$$= \text{ by (6.6) } = d' L^\star + i \, d''^\star.$$

Corollary 6.2.4. *Let* (X, h) *be a Kähler manifold. Then*

$$\boxed{\mathcal{H}_{\Delta}^{l}(X, \mathbb{C}) = \mathcal{H}_{\Delta'}^{l}(X) = \mathcal{H}_{\Delta''}^{l}(X),}$$

$$\boxed{\mathcal{H}_{\Delta'}^{p,q}(X) = \mathcal{H}_{\Delta''}^{p,q}(X).}$$

Set, for $p + q = l$

$$\mathcal{H}_{\Delta}^{p,q}(X) := \mathcal{H}_{\Delta}^{l}(X, \mathbb{C}) \cap A^{p,q}(X).$$

Then

$$\boxed{\mathcal{H}_{\Delta}^{p,q}(X) = \mathcal{H}_{\Delta'}^{p,q}(X) = \mathcal{H}_{\Delta''}^{p,q}(X),}$$

and

$$\boxed{\mathcal{H}_{\Delta}^{l}(X, \mathbb{C}) = \bigoplus_{p+q=l} \mathcal{H}_{\Delta}^{p,q}(X) = \bigoplus_{p+q=l} \mathcal{H}_{\Delta'}^{p,q}(X) = \bigoplus_{p+q=l} \mathcal{H}_{\Delta''}^{p,q}(X).}$$

Proof. It is a consequence of the fact that Δ commutes with $\pi^{p,q}$. The first set of two equalities follow from (6.4). The second set of two equalities from the fact that if u is Δ-harmonic, then it is Δ'-harmonic (Δ''-harmonic, resp.) so that the (p, q)-components of u are Δ'-harmonic (Δ''-harmonic, resp.), hence Δ-harmonic. $\qquad\square$

At this point we could state and prove the Hodge Decomposition Theorem.

However, without some more preparation, it would not be clear from the proof that the resulting decomposition is independent of the Kähler metric. We prefer to prove this most important fact along the way.

6.3   The Hodge Decomposition for compact Kähler manifolds

The following lemma clarifies the relation among the various notions of forms being "closed" in the compact Kähler context. It is used to show that the Hodge decomposition is canonical, i.e. independent of the metric.

Lemma 6.3.1. *(d'd''-**Lemma***) Let* (X, h) *be* <u>*compact Kähler manifold and*</u> $u \in A^{p,q}(X) \cap \operatorname{Ker} d.$

The following are equivalent.

(a) u *is d-exact;*

(b') u *is d'-exact;*

(b'') u *is* d''*-exact;*

(c) u *is* $d'd''$*-exact;*

(d) $u \in (\mathcal{H}^{p,q}(X))^{\perp}$.

Proof. Since $u = d(d''v) = d'(d''v) = d''(-d'v)$, we see that (c) implies (a), (b') and (b'') on any complex manifold.

The rest of the proof needs the compactness as well as the Kähler assumption.

Since (X, h) is Kähler, we have $\Delta' = \Delta'' = \frac{1}{2}\Delta$ so that $\mathcal{H}^{p,q}(X) := \mathcal{H}_{\Delta'}^{p,q}(X) = \mathcal{H}_{\Delta''}^{p,q}(X) = \mathcal{H}_{\Delta}^{p,q}(X)$. In what follows we need the compactness of X to ensure that the inner products are defined. If (a) holds, then $u = dv$ and $\langle\langle dv, w\rangle\rangle = \langle\langle v, d^{\star}w\rangle\rangle = 0$ for every $w \in \mathcal{H}^{p,q}(X)$, since Δ-harmonic forms are characterized by Lemma 2.3.2.

This shows that (a) implies (d).

An analogous argument shows that (b') implies (d) and that (b'') implies (d).

We are left with proving that (d) implies (c).

Since $du = 0$ and u is of pure type (p, q), we have that $d'u = d''u = 0$.

By Theorem 5.2.5 for d'', $u = d''u_1 + d''^{\star}u_2$ for a u_1 of type $(p, q-1)$ and a u_2 of type $(p, q+1)$.

Since $d''u = 0$, we have $d''d''^{\star}u_2 = 0$.

On the other hand, $0 = \langle\langle d''d''^{\star}u_2, u_2\rangle\rangle = ||d''^{\star}u_2||^2$, i.e. $d''^{\star}u_2 = 0$ and $u = d''u_1$.

Theorem 5.2.5 for d' implies that $u_1 = h + d'w_1 + d'^{\star}w_2$, where $h \in \mathcal{H}^{p,q-1}(X)$, w_1 of type $(p-1, q-1)$ and w_2 of type $(p+1, q-1)$.

Using (6.3), we get

$$u = d''u_1 = d''h + d''d'w_1 + d''d'^{\star}w_2 = -d'd''w_1 - d'^{\star}d''w_2.$$

Since $d'u = 0$, the equality above implies that $d'd'^{\star}(d''w_2) = 0$.

The same argument as above, proving that $d''^{\star}u_2 = 0$, shows that $d'^{\star}d''w_2 = 0$, i.e. $u = d'd''(-w_1)$. $\qquad\square$

Theorem 6.3.2. *(The Hodge Decomposition)*

Let X *be a compact Kähler manifold. The natural maps*

$$H_{BC}^{p,q}(X) \longrightarrow H_{dR}^{p+q}(X, \mathbb{C})$$

are injective. We denote their images by $H^{p,q}(X) \subseteq H_{dR}^{p+q}(X, \mathbb{C})$.

There is an internal direct sum decomposition

$$\boxed{H_{dR}^l(X, \mathbb{C}) = \bigoplus_{p+q=l} H^{p,q}(X)}$$

satisfying the equality

$$\boxed{H^{p,q}(X) = \overline{H^{q,p}(X)}.}$$

Proof. Consider the diagram

$$
\begin{array}{ccc}
H^{p,q}_{d''}(X) & \xrightarrow{\ g\ } & \mathcal{H}^{p,q}_{\Delta''}(X) \\
\uparrow f & & \downarrow h \\
H^{p,q}_{BC}(X) & \xrightarrow{\ l\ } & H^{p+q}_{dR}(X,\mathbb{C}).
\end{array}
$$

The maps f and l stem from the definitions. See Exercise 3.7.8. The map g is the inverse of the Hodge Isomorphism Theorem 5.2.5 for d'' and it selects the Δ''-harmonic representative for a Dolbeault class. The map h is defined by the fact that a Δ''-harmonic class is Δ-harmonic, hence d-closed.

CLAIM I: *f is bijective.* Injectivity follows from Lemma 6.3.1, $(b'') = (c)$. Surjectivity follows from the fact that every element $a_D \in H^{p,q}_{d''}(X)$ can be represented by a Δ''-harmonic (p,q)-form a which is then necessarily d-closed so that the associated class $a_{BC} \in H^{p,q}_{BC}(X)$ maps to a_D.

CLAIM II: *the diagram commutes.* Let $a \in A^{p,q}(X) \cap \operatorname{Ker} d$. Let $a_{BC} \in H^{p,q}_{BC}(X)$, $a_D \in H^{p,q}_{d''}(X)$ and $a_{dR} \in H^{p+q}_{dR}(X,\mathbb{C})$ be the corresponding elements. Let $a' \in \mathcal{H}^{p,q}_{\Delta''}(X)$ be the Δ''-harmonic representative of a_D and $a'_{dR} \in H^{p+q}_{dR}(X,\mathbb{C})$ be the corresponding class. We need to show that $a_{dR} = a'_{dR}$, i.e. that $a - a'$ is d-exact. Note that $a' \in A^{p,q}(X) \cap \operatorname{Ker} d$. We have that $f((a - a')_{BC}) = 0$. By CLAIM I, $(a - a')$ is $d'd''$-exact. We conclude by Lemma 6.3.1, $(a) = (c)$.

By Corollary 6.2.4, h is injective. It follows that l is injective and that the image of $\mathcal{H}^{p,q}_{\Delta''}(X)$ in $H^{p+q}_{dR}(X,\mathbb{C})$, being the image of $H^{p,q}_{BC}(X)$ which depends only on the complex structure of X, is independent of the Kähler metric employed.

By Corollary 6.2.4, we can assemble, for all pair of indices (p,q) such that $l = p + q$, the isomorphisms l and reach the desired conclusion. $\qquad \square$

Remark 6.3.3. (The meaning of the Hodge Decomposition) The Hodge Decomposition expresses a basic and important property of the complex cohomology of compact Kähler manifolds. It does *not* state that any complex-valued de Rham class $a_{dR} \in H^{p+q}_{dR}(X,\mathbb{C})$ admits a canonical representative $a \in A^l(X)$ and associated (p,q)-components. It states that a_{dR} can be canonically decomposed into a sum $\sum a^{p,q}_{dR}$, whose terms one calls the (p,q)-components of a_{dR} and which do not depend on the choice of a Kähler metric. The choice of a Kähler metric allows to represent these (p, q)-components using Δ-harmonic (p, q)-forms which depend on the metric.

This canonical decomposition of $H^l_{dR}(X, \mathbb{C})$ and the canonical isomorphism $H^l_{dR}(X, \mathbb{C}) \simeq H^l(X, \mathbb{Z}) \otimes_{\mathbb{Z}} \mathbb{C}$ lead to the notion of pure Hodge structure of weight l. See §7.1.

Remark 6.3.4. For a general complex manifold X, there is no natural map $H^{p,q}_{d''}(X) \longrightarrow H^{p+q}_{dR}(X, \mathbb{C})$: the map $h \circ g$ depends on the choice of a metric and f is not, in general, invertible.

In the compact Kähler case, we have shown that $k := h \circ g = l \circ f^{-1}$ is well-defined and depends only on the complex structure on X and we have

$$H^{p,q}(X) := l(H^{p,q}_{BC}(X)) = k(H^{p,q}_{d''}(X)) = h(\mathcal{H}^{p,q}_{\Delta''}(X)).$$

6.4 Some consequences

The following Theorem is trivially false on a complex manifold and the example of the Iwasawa manifold Example 6.1.3 shows that it is false in general on compact complex manifolds. However, it is true on compact complex surfaces.

Theorem 6.4.1. *(Holomorphic vs. closed forms)* Let X be a *compact* Kähler manifold and $u \in H^0(X, \Omega^p_X)$ be a holomorphic p-form, i.e. a $(p, 0)$-form such that $d''u = 0$.

Then u is closed.

Proof. Let h be a Kähler metric on X. Since $d''^* u$ has bi-degree $(p, -1)$, it must be the zero form. It follows that u is Δ''-harmonic, hence Δ-harmonic, hence d-closed. (See the remark after the Proof of Lemma 2.3.2.) $\quad\square$

Let us list some of the properties we have proved concerning the *Hodge numbers*

$$h^{p,q}(X) := \dim_{\mathbb{C}} H^q(X, \Omega^p_X) = \dim_{\mathbb{C}} H^{p,q}_{d''}(X) = \dim_{\mathbb{C}} H^{q,p}_{d'}(X)$$

for compact complex manifolds and for compact Kähler manifolds. Note that we do not have $h^{p,q}(X) = h^{q,p}(X)$, unless X is Kähler, e.g. the Hopf surface for which $h^{0,1}(X) = h^{1,0}(X) + 1 = 1$. Note that in this case the so-called Hodge-Frölicher spectral sequence collapses, as it does for any compact complex surface, so that even this last condition does not imply $h^{p,q}(X) = h^{q,p}(X)$. It is only equivalent to $b_l(X) = \sum_{p+q=l} h^{p,q}(X)$.

In what follows we omit the Künneth-type relations. See [Griffiths and Harris, 1978], p. 105.

Theorem 6.4.2. *Let X be a compact Kähler manifold of dimension n. We have*

$$h^{p,q}(X) < \infty, \tag{6.18}$$

$$h^{0,0}(X) = h^{n,n}(X) = 1, \tag{6.19}$$

$$h^{p,q} = h^{n-p,n-q}(X). \tag{6.20}$$

Assume in addition that X is Kähler. Then we also have

$$h^{p,q}(X) = h^{q,p}(X), \tag{6.21}$$

$$b_l(X) = \sum_{p+q=l} h^{p,q}(X). \tag{6.22}$$

In particular, $b_{odd}(X) = even$.

$$h^{p,p}(X) \geq 1, \qquad 0 \leq p \leq n. \tag{6.23}$$

Proof. Exercise.   □

These relations can be visualized conveniently on the so-called *Hodge diamond* [Griffiths and Harris, 1978], p. 117.

There are also some inequalities stemming from the Hard Lefschetz Theorem 7.3.4. See [Demailly, 1996], Corollaire 8.17 for the precise statements, which are immediate consequences of Theorem 7.3.4.b.

Exercise 6.4.3. Compute $h^{p,q}(\mathbb{P}^n)$.

Exercise 6.4.4. Compute $h^{p,q}(Q^n)$, where $Q^n \subseteq \mathbb{P}^{n+1}$ is the nonsingular n-dimensional quadric.

Lecture 7

The Hard Lefschetz Theorem and the Hodge-Riemann Bilinear Relations

We discuss Hodge structures and polarizations, the operation of cupping with the fundamental class of a hyperplane section, the Hard Lefschetz Theorem, the Hodge-Riemann Bilinear Relations and the Weak Lefschetz Theorem (which is also called the Lefschetz Theorem on Hyperplane Sections).

The statement of the Hard Lefschetz Theorem for a compact Kähler manifold (X, ω) of dimension n is perhaps surprising at first.

Poincaré Duality asserts that
$$H^{n-j}(X, \mathbb{C}) \simeq H^{n+j}(X, \mathbb{C})^{\vee}.$$
In particular, $b_{n-j}(X) = b_{n+j}(X)$.

The Hard Lefschetz Theorem states that
$$L^j : H^{n-j}(X, \mathbb{C}) \longrightarrow H^{n+j}(X, \mathbb{C}), \qquad u \longrightarrow \omega^j \wedge u$$
is an isomorphism.

This isomorphism, coupled with Poincaré Duality, gives rise to the non-degenerate bilinear form
$$H^{n-j}(X, \mathbb{C}) \times H^{n-j}(X, \mathbb{C}) \longrightarrow \mathbb{C}, \qquad (u, v) \longrightarrow \int_X \omega^j \wedge u \wedge v.$$
The Hodge-Riemann Bilinear Relations express the beautiful signature properties of this bilinear form.

Though the Hard Lefschetz Theorem and the Hodge Riemann Bilinear relations hold for any compact Kähler manifold (X, ω), in the last two lectures we discuss the projective case (a projective manifold is compact and Kähler) which has some additional geometric flavor with respect to the compact Kähler case: one can choose ω to be the fundamental class of a hyperplane section on X.

71

7.1   Hodge structures

The theory of Hodge structures and polarizations is a powerful tool to investigate the topology of algebraic varieties. It has been employed fruitfully to study families of smooth projective varieties (variations of polarized pure Hodge structures).

The fact, discovered by Deligne, that the singular cohomology of a complex algebraic variety is endowed with a canonical and functorial mixed Hodge structure is one of beautiful depth.

The Hodge Decomposition Theorem and the Hodge-Riemann Bilinear Relations can be conveniently re-formulated in terms of Hodge structures and polarizations.

Let $l \in \mathbb{Z}$, H be a finitely generated abelian group, $H_{\mathbb{Q}} := H \otimes_{\mathbb{Z}} \mathbb{Q}$, $H_{\mathbb{R}} = H \otimes_{\mathbb{Z}} \mathbb{R}$, $H_{\mathbb{C}} = H \otimes_{\mathbb{Z}} \mathbb{C}$.

Definition 7.1.1. (Hodge structure) A *pure Hodge structure of weight l on H,* or on $H_{\mathbb{Q}}$, $H_{\mathbb{R}}$, is a direct sum decomposition

$$\boxed{H_{\mathbb{C}} = \bigoplus_{p+q=l} H^{p,q}}$$

such that

$$\boxed{H^{p,q} = \overline{H^{q,p}}.}$$

One can change the labeling of the $H^{p,q}$ spaces and change the weight l to any other weight l' with $l' \equiv l \, mod \, 2$.

The *Hodge filtration* on such a structure is the <u>decreasing</u> filtration

$$F^p(H_{\mathbb{C}}) := \bigoplus_{p' \geq p} H^{p', l-p'}.$$

A *morphism of pure Hodge structures* $f : H \longrightarrow H'$ of the same weight l is a group homomorphism such that $f \otimes Id_{\mathbb{C}}$ is compatible with the Hodge filtration, i.e. such that it is a filtered map, i.e. such that

$$f(F^p(H_{\mathbb{C}})) \subseteq F^p(H'_{\mathbb{C}}).$$

Such maps are automatically *strict* in the sense that
$$\text{Im}(F^p(H_{\mathbb{C}})) = \text{Im}(H_{\mathbb{C}}) \cap F^p(H'_{\mathbb{C}}).$$
The category of pure Hodge structures of weight l is abelian. In particular, kernels and cokernels of maps of pure Hodge structures of weight l are pure Hodge structures of weight l.

Exercise 7.1.2. Give examples of filtered maps of vector spaces which are not strict.

Example 7.1.3. With reference to §4.1, the vector space $\Lambda^l_{\mathbb{C}}(V^*_{\mathbb{C}})$ is a pure Hodge structure of weight l. If $f : X \to Y$ is a holomorphic map of complex manifolds, then, for every $x \in X$, the induced map
$$f^* : \Lambda^l_{\mathbb{C}}(T^*_{Y,f(x)}(\mathbb{C})) \longrightarrow \Lambda^l_{\mathbb{C}}(T^*_{X,x}(\mathbb{C}))$$
is a map of Hodge structures.

Example 7.1.4. The Hodge Decomposition Theorem 6.3.2 can be reformulated by stating that if X is a compact Kähler manifold, then $H^l(X, \mathbb{Z})$ admits a natural structure of a pure Hodge structure of weight l. The reader should check that this structure is functorial with respect to holomorphic maps of compact Kähler manifolds.

Exercise 7.1.5. Verify the statements of Examples 7.1.3 and 7.1.4.

Exercise 7.1.6. Let X be compact Kähler and L be a complex line bundle on X. Denote its first Chern class and the operation of cupping with this first Chern class simply by L.

Observe that L has bidegree $(1,1)$ as an operator on $H^{\bullet}(X, \mathbb{C})$ in the sense that it maps a class of degree (p,q) to one of degree $(p+1, q+1)$.

Re-define the Hodge structure of weight $l + 2r$ on $H^{l+2r}(X, \mathbb{Z})$ so that $L^r : H^l(X, \mathbb{Z}) \longrightarrow H^{l+2r}(X, \mathbb{Z})$ becomes a morphism of Hodge structures of weight l. Do the analogous operation for weight $l + 2$ and for any weight $l + 2m$, $m \in \mathbb{Z}$.

Show directly that $\text{Ker}\, L^r$, $\text{Im}\, L^r$ and $\text{Coker}\, L^r$ are pure Hodge structures of weight l.

Let C be the Weil operator, i.e.

$$C : H_{\mathbb{C}} \simeq H_{\mathbb{C}}, \qquad C(x) = i^{p-q}x, \quad \forall x \in H^{p,q}.$$

See §4.5. The Weil operator is real, i.e. it fixes $H_{\mathbb{R}}$.

Replacing i^{p-q} by $z^p \bar{z}^q$ we get a real action ρ of $\mathbb{C}^*$ on $H_{\mathbb{R}}$ and $C = \rho(i)$.

Definition 7.1.7. A *polarization* of the real pure Hodge structure $H_{\mathbb{R}}$ of weight l is a real bilinear form Ψ on $H_{\mathbb{R}}$ which is invariant under the action given by ρ restricted to $S^1 \subseteq \mathbb{C}^*$ and such that the bilinear form

$$\boxed{\widetilde{\Psi}(x,y) := \Psi(x, C(y))}$$

is symmetric and positive definite.

Exercise 7.1.8. Verify the following statements. If Ψ is a polarization of the real pure Hodge structure $H_{\mathbb{R}}$ of weight l, then Ψ is symmetric if l is even, and antisymmetric if l is odd. (Hint: $C^2 = (-1)^l$.)

In any case, Ψ is non-degenerate.

In addition, for every $0 \neq x \in H^{p,q}$, $(-1)^l i^{p-q} \Psi(x, \bar{x}) > 0$, where, by abuse of notation, Ψ also denotes the $\mathbb{C}$-bilinear extension of Ψ to $H_{\mathbb{C}}$. (Hint: consider $x + \bar{x}$ and $\frac{x - \bar{x}}{i}$)

Remark 7.1.9. If $H' \subseteq H$ is a Hodge sub-structure, then $H'_{\mathbb{R}}$ is fixed by C so that $\Psi_{|H'_{\mathbb{R}}}$ is a polarization.

In particular, $\Psi_{|H'_{\mathbb{R}}}$ is non-degenerate.

7.2   The cup product with the Chern class of a hyperplane bundle

Let M be a smooth manifold. An l-form $\omega \in E^l(M)$ defines a linear map

$$L : E^\bullet(M) \longrightarrow E^{\bullet+l}(M) \qquad u \longrightarrow \omega \wedge u.$$

If ω is closed, then L descends to de Rham cohomology:

$$L : H^\bullet_{dR}(M, \mathbb{R}) \longrightarrow H^{\bullet+l}_{dR}(M, \mathbb{R}), \qquad u \longrightarrow \omega \wedge u.$$

Since the de Rham isomorphism $H^\bullet_{dR}(M, \mathbb{R}) \simeq H^\bullet(M, \mathbb{R})$ is an algebra isomorphism, we may write $u \cup b$ or $u \wedge v$ depending on our taste and notational convenience.

If, in addition, M is compact and oriented of dimension m, then we also have the Poincaré Duality isomorphism which we denote

$$H^\bullet(M, \mathbb{R}) \overset{PD}{\simeq} H_{m-\bullet}(M, \mathbb{R}).$$

We denote the inverse map by the same symbol.

We have the following geometric interpretation of the cap, cup (wedge) products in connection with set-theoretic intersections.

The *cap* product

$$\cap : H^q(M, \mathbb{Z}) \otimes_{\mathbb{Z}} H_p(M, \mathbb{Z}) \longrightarrow H_{p-q}(M, \mathbb{Z})$$

satisfies, taking $\mathbb{Q}$-coefficients, denoting cohomology classes by Roman letters and homology classes by Greek ones:

$$\boxed{a \cap \beta = (a \cup \beta^{PD})^{PD} = a^{PD} \cap \beta,}$$

where $a^{PD} \cap \beta$ is the result of the following operations: pass from a to a^{PD}; select representatives for a^{PD} and for β so that their supports meet transversally; the correctly oriented intersection is a well-defined homology class.

In other words: *the cap/cup products correspond to transverse intersections via Poincaré Duality.*

Exercise 7.2.1. Let L be the y-axis in $\mathbb{R}^2$ oriented by dy and C be the unit circle centered about the origin oriented counterclockwise.

Verify that the intersection numbers, with respect to the standard orientation for $\mathbb{R}^2$, are $i_{(0,1)}(L \cap C) = -1$ and $i_{(0,-1)}(L \cap C) = 1$ so that the total intersection number is zero.

Do the analogous thing in $\mathbb{C}^2$ and verify that the total intersection number is 2.

Verify that if two complex varieties of complementary dimensions in $\mathbb{C}^n$ meet transversally at a point, then the intersection number at that point is positive.

In fact, this is true even if they do not meet transversally but still meet at isolated points.

See [Griffiths and Harris, 1978], pp. 49-65 for a nice discussion on intersecting cycles.

Remark 7.2.2. (Borel-Moore Homology) Two distinct lines L_1 and L_2 in $\mathbb{P}^2_{\mathbb{C}}$ meet at one point, say P, with intersection index one.

Singular cohomology records this fact as follows.

The lines give rise to cohomology classes $cl(L_i) \in H^2(\mathbb{CP}^2, \mathbb{Z})$ and we get

$$\deg\left(cl(L_1) \cup cl(L_2)\right) = 1$$

where $\deg : H^4(\mathbb{CP}^2, \mathbb{Z}) \simeq \mathbb{Z}$ is the isomorphism stemming from the standard orientation of $\mathbb{P}^2$.

Let $Q \in \mathbb{P}^2$ be a point on neither line. We have $H^4(\mathbb{CP}^2 \setminus Q, \mathbb{Z}) = \{0\}$. The pairing in cohomology is not the correct one to deal with intersections in a non-compact ambient manifold.

Moreover, if we consider non-compact subvarieties and their intersections, e.g. two lines in $\mathbb{C}^2$ meeting transversally at a point, we see that the non-compactness of the subvarieties also plays a role in the loss of information

It turns out that the formalism of Borel-Moore Homology, with its natural pairing with singular homology, allows for a reasonably general intersection theory. Given two complex subvarieties V and W of a complex manifold X of dimension n we have a paring, depending on X,

$$H_j^{BM}(V, \mathbb{Q}) \times H_k(W, \mathbb{Q}) \longrightarrow H_{2n-j-k}^{BM}(V \cap W, \mathbb{Q})$$

with a natural degree map giving intersection numbers if $j + k = 2n$.

The cycles in the Borel-Moore theory are locally finite combinations of simplices so that it is natural to expect a pairing with the cycles is homology which are given by a finite number of simplices.

7.3 The Hard Lefschetz Theorem and the Hodge-Riemann Bilinear Relations

Let $X \subseteq \mathbb{P}^N$ be a nonsingular complex projective variety not contained in any hyperplane.

The space of holomorphic sections of the line bundle

$$\boxed{L := L_{1|X},}$$

where L_1 is the hyperplane line bundle of Exercise 3.2.1, contains a subspace $\mathcal{L}$ naturally identified with the space $\mathbb{C}[x_0, \ldots, x_N]_1$ of homogeneous linear polynomials in $N + 1$ variables.

If $s \in \mathcal{L} \setminus \{0\}$ and P_s is the corresponding polynomial, then, counting multiplicities, the zero sets

$$\{s = 0\} = \{X \cap \{P_s = 0\}\}.$$

We have the Fubini-Study metric h_{FS} on $\mathbb{P}^N$ with associated Kähler form ω_{FS}.

Restricting to X we get a Kähler metric h with associated Kähler form $\omega := \omega_{FS|X}$ and we have, by slight abuse of notation:

$$\omega = (X \cap \mathbb{P}^{N-1})^{PD} = c_1(L) \in H^2(X, \mathbb{Q}).$$

It follows, via Poincaré Duality, that

---

cupping with ω corresponds to intersecting with a hyperplane section.

Exercise 7.3.1. Let $C \subseteq \mathbb{P}^N$ be a nonsingular complex projective curve. Show that

$$L^r : H^{1-r}(C, \mathbb{Q}) \simeq H^{1+r}(C, \mathbb{Q}), \quad \forall r \geq 0.$$

Show that

$$\int_C \omega \wedge u \wedge \overline{u} > 0, \quad \forall u \in H^{0,0}(C).$$

$$i^{p-q} \int_C u \wedge \overline{u} > 0, \quad \forall u \in H^{p,q}(C), \ p+q = 1.$$

Exercise 7.3.2. Let $S \subseteq \mathbb{P}^N$ be a nonsingular complex algebraic surface. Show that

$$L^r : H^{2-r}(C, \mathbb{Q}) \simeq H^{2+r}(C, \mathbb{Q}), \quad r = 0, 2.$$

Deduce that

$$H^2(S, \mathbb{Q}) = L(H^0(S, \mathbb{Q})) \overset{\perp}{\oplus} \mathrm{Ker} L$$

where the direct sum is orthogonal with respect to the natural cup product pairing on $H^2(S, \mathbb{Q})$, each direct summand is a pure Hodge structure of weight 2 and $\dim_{\mathbb{Q}} \mathrm{Ker} L = b_2 - b_0 = b_2 - 1$.

Let $S \subseteq \mathbb{P}^3$ be a nonsingular quadric. Show that the restriction of the cup product to $\mathrm{Ker} L \subseteq H^2(S, \mathbb{Q})$ is negative definite. Show that the same is true no matter which ample line bundle L one chooses on S.

The methods of Exercise 7.3.2 do not seem to be helpful in studying the map

$$L \, : \, H^1(S, \mathbb{Q}) \longrightarrow H^3(S, \mathbb{Q}).$$

Perhaps, at this point, the reader unfamiliar with the Hard Lefschetz Theorem 7.3.4.a will appreciate its beauty.

Let $0 \le r \le n$ and define the *space P_L^{n-r} of rational $(n-r)$-primitive classes* as

$$P^{n-r} := \operatorname{Ker} L^{r+1} \subseteq H^{n-r}(X, \mathbb{Q}).$$

Clearly, this space can be defined also for $\mathbb{Z}$, $\mathbb{R}$ and $\mathbb{C}$ coefficients, in which case we use the notation $P_L^{n-r}(X, \mathbb{R})$, for example.

Exercise 7.3.3. Show that

$$P_L^j(X, \mathbb{C}) = \bigoplus_{p+q=j} (\, P_L^j(X, \mathbb{C}) \cap H^{p,q}(X)\,), \quad 0 \le j \le n,$$

i.e. show that $P_L^j \subseteq H^j(X, \mathbb{Q})$ is a pure sub-Hodge structure of weight j.

Theorem 7.3.4. *(a) (***The Hard Lefschetz Theorem.***) Let $X \subseteq \mathbb{P}^N$ be a projective manifold of dimension n.*
For every $r \ge 0$, we have isomorphisms

$$L^r \, : \, H^{n-r}(X, \mathbb{Q}) \longrightarrow H^{n+r}(X, \mathbb{Q}).$$

*(b) (***The primitive Lefschetz decomposition.***) For every $r \ge 0$ there is the direct sum decomposition*

$$H^{n-r}(X, \mathbb{Q}) = \bigoplus_{j \ge 0}^{\perp} L^j P_L^{n-r-2j}$$

where each summand is a pure Hodge sub-structure of weight $n - r$ and all summands are mutually orthogonal with respect to the bilinear form $\int_X \omega^r \wedge - \wedge -$.

*(c) (***The Hodge-Riemann bilinear relations.***) For every $0 \le l \le n$, the bilinear form $(-1)^{\frac{l(l+1)}{2}} \int_X \omega^r \wedge - \wedge -$ is a polarization of the real pure weight l Hodge structure $P_L^l(X, \mathbb{R}) \subseteq H^l(X, \mathbb{R})$. In particular,*

$$(-1)^{\frac{l(l-1)}{2}} i^{p-q} \int_X \omega^{n-l} \wedge \alpha \wedge \overline{\alpha} > 0, \qquad \forall \, 0 \ne \alpha \in P_L^l(X, \mathbb{C}) \cap H^{p,q}(X, \mathbb{C}).$$

Remark 7.3.5. Theorem 7.3.4.a was first stated by Lefschetz in his 1924 book. It is known that his "proof" contains a gap. The first proof is due to Hodge and uses what is known today as "Hodge Theory." For a proof based on the commutation relations and how they relate to a $sl_2(\mathbb{C})$-action on $H^\bullet(X, \mathbb{C})$ see [Griffiths and Harris, 1978], pp. 118-122, or [Demailly, 1996], pp. 35-39 and 44-45. A proof of Theorem 7.3.4.c can be found in [Weil, 1958], [Wells, 1980], or [Voisin, 2002].

In his second paper on the Weil conjectures, Deligne has given an algebraic proof of Theorem 7.3.4.a. It uses algebraic geometry in characteristic p. See §8.5 for a discussion. It seems unlikely that those methods will give a proof of Theorem 7.3.4.c.

Remark 7.3.6. In view of Remark 7.1.9, the bilinear form Ψ, which is non-degenerate on the primitive spaces, is in fact non-degenerate when restricted to any sub-Hodge structure of the primitive spaces.

Exercise 7.3.7. Show that Theorem 7.3.4.a is equivalent to the statement that $L^j : H^{n-r}(X, \mathbb{Q}) \longrightarrow H^{n-r+2j}(X, \mathbb{Q})$ is injective for $0 \le j \le r$.

Show that Theorem 7.3.4.a is equivalent to the statement that $L^j : H^{n+r-2j}(X, \mathbb{Q}) \longrightarrow H^{n+r}(X, \mathbb{Q})$ is surjective for $0 \le j \le r$.

Show that Theorem 7.3.4.a implies Theorem 7.3.4.b.

Exercise 7.3.8. Show that Theorem 7.3.4.a is equivalent to the bilinear form

$$\Psi(u, v) := \int_X \omega^r \wedge u \wedge v$$

on $H^{n-r}(X, \mathbb{Q})$ being non-degenerate.

Exercise 7.3.9. Prove the Hard Lefschetz Theorem when $X = \mathbb{P}^n$, using only the product structure on the cohomology ring.

Example 7.3.10. Let $X \longrightarrow Y$ be the blowing-up of a point $y \in Y$ on a smooth projective three-fold Y, A be an ample line bundle on Y and $L := f^*A$. Then the statement of the Hard Lefschetz Theorem 7.3.4.a does *not* hold for L. In fact, if $D = f^{-1}(y)$ is the exceptional divisor, then $L \cup [D] = 0$ so that $L : H^{3-1}(X, \mathbb{Q}) \longrightarrow H^{3+1}(X, \mathbb{Q})$ is not an isomorphism.

Remark 7.3.11. What follows is proved in [de Cataldo and Migliorini, 2002]. Let $f : X \longrightarrow Y$ be a map of complex projective varieties, X nonsingular, A ample on Y, $L := f^*A$.

Then the statement of the Hard Lefschetz Theorem 7.3.4.a holds for L *if and only if* the map f is *semi-small*. The notion of semi-smallness was introduced by Goresky and MacPherson to generalize the Lefschetz Theorem on Hyperplane Sections 7.4.1 to the case of not necessarily proper complex analytic maps $f : X \to \mathbb{P}^N$. A map $f : X \longrightarrow Y$ is semi-small iff $\dim X \times_Y X = \dim X$. Note that in general, we have the inequality "$\geq$." These maps occur frequently in complex geometry and in representation theory. If f is semi-small, then one also shows that all three statements of Theorem 7.3.4 hold for f.

7.4   The Weak Lefschetz Theorem

The following is another fundamental result about the topology of projective varieties. Essentially, it states that a great deal of the topology of a projective manifold "comes" from its hyperplane sections.

Let $Y := X \cap \mathbb{P}^{N-1} \subseteq X \subseteq \mathbb{P}^N$ be a hyperplane section of a projective manifold X of complex dimension n embedded in projective space.

Theorem 7.4.1. *(The Weak Lefschetz Theorem.)* *The natural restriction map*

$$r^* : H^j(X, \mathbb{Q}) \longrightarrow H^j(Y, \mathbb{Q})$$

is an isomorphism for $j \leq n - 2$ and is injective for $j = n - 1$.

If, in addition, Y is nonsingular, then the natural Gysin map (i.e. the Poincaré dual to the map in homology)

$$\tilde{r}_* : H^{n+j-2}(Y) \longrightarrow H^{n+j}(X)$$

is an isomorphism for $j \geq 2$ and is surjective for $j = 1$.

Sketch of proof. For a proof and other references, see [Griffiths and Harris, 1978], p. 156.

There is an analogus result for the homotopy groups, but we do not state it here.

The complex manifold $X \setminus Y$ is a closed complex submanifold of $\mathbb{C}^N$ of dimension n.

Andreotti and Frankel have observed that such a manifold has the homotopy type of a CW-complex of real dimension at most n so that

$$H^j(X \setminus Y, \mathbb{Q}) = \{0\}, \qquad \forall\, j > n.$$

Equivalently,

$$H_c^j(X \setminus Y, \mathbb{Q}) = \{0\}, \qquad \forall\, j < n.$$

To conclude one uses the long exact sequence of relative cohomology

$$\longrightarrow H_c^j(X \setminus Y, \mathbb{Q}) \longrightarrow H^j(X, \mathbb{Q}) \longrightarrow H^j(Y, \mathbb{Q}) \longrightarrow H_c^{j+1}(X \setminus Y, \mathbb{Q}) \longrightarrow .$$

$\square$

Exercise 7.4.2. C.P. Ramanujan has elegantly shown that the Weak Lefschetz Theorem and the Kodaira Vanishing Theorem, i.e. $H^j(X, L^\vee) = \{0\}$ for $j < n$, are equivalent.

Prove this fact by looking at the long exact sequences stemming from the following two exact sequences:

$$0 \longrightarrow \Omega_X^p \otimes L^\vee \longrightarrow \Omega_X^p \overset{r}{\longrightarrow} \Omega_{X|Y}^p \longrightarrow 0,$$

$$0 \longrightarrow \Omega_Y^{p-1} \otimes L_{|Y}^\vee \longrightarrow \Omega_{X|Y}^p \overset{i}{\longrightarrow} \Omega_Y^p \longrightarrow 0.$$

You will need the functoriality of the Hodge Decomposition, the Dolbeault isomorphism, and the fact $i \circ r$ induces the restriction of p-forms from X to Y. See [Griffiths and Harris, 1978], p. 157.

Theorem 7.4.1 imposes severe restrictions on the topology of algebraic varieties. For example, it implies that if X is a surface then the natural map of fundamental groups

$$\pi_1(Y, y) \longrightarrow \pi_1(X, y)$$

is surjective. If X is not simply connected, then a smooth hyperplane section Y cannot be isomorphic to $\mathbb{P}^1$.

If X is simply connected and it has dimension at least three, then a smooth hyperplane section Y must be simply connected and the natural restriction map

$$Pic(X) \longrightarrow Pic(Y)$$

of Picard groups (isomorphism classes of holomorphic line bundles with product induced by the tensor product) is injective with torsion free-cokernel and if the dimension of X is at least four, then the restriction map is an isomorphism, i.e. every line bundle on Y comes from one on X by restriction (pull-back).

Mixed Hodge structures, semi-simplicity and approximability

We discuss the mixed Hodge structure on the singular cohomology of a complex algebraic variety, the Semi-simplicity Theorem, the Global Invariant Cycle Theorem, the connection between the Semi-simplicity Theorem and the Hard and Weak Lefschetz Theorems. We conclude with illustrating an approximation technique for primitive vectors.

8.1 The mixed Hodge structure on the cohomology of complex algebraic varieties

The paper [Durfee, 1981] is a nice introduction to mixed Hodge structures.

The following is Deligne's fundamental result which endows in a functorial way the singular cohomology of a complex algebraic variety with a mixed Hodge structure. See [Deligne, 1971] and [Deligne, 1974].

The implications of the existence of this rich structure are numerous and far-reaching, for they impose severe constraints of varieties and the maps between them. We try to get a glimpse of this powerful machinery in this last lecture.

The following summarizes some of the basic properties enjoyed by the cohomology of complex algebraic varieties.

Theorem 8.1.1. *Let X be an algebraic variety. For each j there is an increasing weight filtration*

$$\boxed{\{0\} = W_{-1} \subseteq W_0 \subseteq \ldots \subseteq W_{2j} = H^j(X, \mathbb{Q})}$$

and a decreasing Hodge filtration

$$H^j(X, \mathbb{C}) = F^0 \supseteq F^1 \supseteq \ldots \supseteq F^m \supseteq F^{m+1} = \{0\}$$

such that the filtration induced by $F^\bullet$ on the complexified graded pieces of the weight filtration endows every graded piece W_l/W_{l-1} with a pure Hodge structure of weight l.

This structure is functorial for maps of algebraic varieties and the induced maps strictly preserve both filtrations.

Here is a very partial list of the properties of this structure. See [Durfee, 1981] and [Deligne, 1974], §8.2 for some more properties.

- If X is projective and smooth, then

$$0 = W_{j-1} \subseteq W_j = H^j(X, \mathbb{Q})$$

  and one says that $H^j(X, \mathbb{Q})$ *has pure weight j.*
- If X is projective (but not necessarily smooth), then

$$W_j = H^j(X, \mathbb{Q})$$

  and one says that $H^j(X, \mathbb{Q})$ *has weights $\leq j$.*
- If X is smooth (but not necessarily projective), then

$$\{0\} = W_{j-1} \subseteq H^j(X, \mathbb{Q})$$

  and one says that $H^j(X, \mathbb{Q})$ *has weights $\geq j$.*
- If $i : U \hookrightarrow X$ is the inclusion of a Zariski-dense open subset of a smooth projective manifold, then

$$W_j(H^j(U, \mathbb{Q})) = i^* H^j(X, \mathbb{Q}) = i^* W_j(H^j(X, \mathbb{Q})).$$

A beautiful application of the theory of mixed Hodge structures is the following theorem of Deligne's whose proof is a nice and simple exemplification of the "yoga of weights."

Theorem 8.1.2. *Let $Y \longrightarrow U \hookrightarrow X$ be maps of complex algebraic varieties, with Y proper, X proper and nonsingular, U a Zariski-dense open subset of X.*

Then the natural images of $H^j(X, \mathbb{Q})$ and of $H^j(U, \mathbb{Q})$ in $H^j(Y, \mathbb{Q})$ coincide.

Proof. See [Deligne, 1971], Corollaire 3.2.18 and [Deligne, 1974], Proposition 8.2.6.

By strictness, we have

$$\operatorname{Im}(H^j(U,\mathbb{Q})) \cap W_j(H^j(Y,\mathbb{Q})) = \operatorname{Im}(W_j(H^j(U,\mathbb{Q}))).$$

Since $H^j(Y,\mathbb{Q})$ has weights $\leq j$, we have that

$$\operatorname{Im}(H^j(U,\mathbb{Q})) = \operatorname{Im}(W_j(H^j(U,\mathbb{Q}))).$$

Since $W_j(H^j(U,\mathbb{Q})) = i^* H^j(X,\mathbb{Q})$ we get the desired equality. $\square$

Example 8.1.3. Let $S^1 \longrightarrow \mathbb{R}^2 \setminus \{0\} \subseteq S^2$ be the obvious inclusion of real algebraic varieties. The conclusion of Theorem 8.1.2 does *not* hold in this situation.

Theorem 8.1.2 expresses another property peculiar to *complex* algebraic geometry.

8.2 The Semi-simplicity Theorem

A *local system* on a (connected) variety Y is a locally constant sheaf $\mathcal{L}$ of finite dimensional rational vector spaces on Y.

Let $\mathcal{L}$ be a local system on Y, $y \in Y$ and $\mathcal{L}_y$ the stalk at y.

The vector space $\mathcal{L}_y$ is a representation of the fundamental group $\pi_1(Y,y)$ and the sheaf $\mathcal{L}$ is the sheaf of sections of the vector bundle $\widetilde{Y} \times_{\pi_1(Y,y)} \mathcal{L}_y$, where $\widetilde{Y}$ is the universal cover of Y.

From this description one deduces a canonical isomorphism

$$H^0(Y,\mathcal{L}) \simeq \mathcal{L}_y^{\pi_1(Y,y)} \tag{8.1}$$

where the right-hand side denotes the subspace of $\mathcal{L}_y$ of vectors fixed by $\pi_1(Y,y)$.

Example 8.2.1. Let $f : X \longrightarrow Y$ be a proper, surjective and smooth map of smooth manifolds such that df has everywhere maximal rank.

By Ehreshmann Lemma, [Demailly, 1996], Lemme 10.2, f is locally topologically trivial over Y, i.e. all fibers are diffeomorphic and, locally over Y, X is a product of Y times the typical fiber.

Circuiting around closed paths centered at a point $y \in Y$ produces a local system on Y by considering the action of $\pi_1(Y, y)$ on the cohomology groups $H^j(f^{-1}(y), \mathbb{Q})$ of the fiber at y.

The sheaf on Y so obtained is canonically isomorphic to the j-th higher direct image $R^j f_* \mathbb{Q}_X$ of the constant sheaf $\mathbb{Q}_X$.

Definition 8.2.2. A local system $\mathcal{L}$ on an algebraic variety Y is said to be *semi-simple* if every local subsystem $\mathcal{L}'$ of $\mathcal{L}$ admits a complement, i.e. a local subsystem $\mathcal{L}''$ of $\mathcal{L}$ such that

$$\mathcal{L} \simeq \mathcal{L}' \oplus \mathcal{L}''.$$

Example 8.2.3. The local system $\mathcal{L}$ on the circle S^1 given by the representation

$$\rho : \pi_1(S^1, q) \simeq \mathbb{Z} \longrightarrow GL(2, \mathbb{Q}), \qquad \rho(1) := \begin{pmatrix} 1 & 1 \\ 0 & 1 \end{pmatrix}$$

is indecomposable, i.e. it cannot be written as a direct sum of non-trivial local sub-systems, but it is *not* semi-simple since it admits the local subsystem generated by the column vector $(1, 0)$.

This local system appears in connection with the standard degeneration over the unit disk $\Delta \subseteq \mathbb{C}$ of a smooth elliptic curve to a rational nodal curve.

The following result of Deligne's summarizes some very important properties of smooth projective maps.

Theorem 8.2.4. *(The Semi-simplicity Theorem for smooth maps)* Let $f : X^n \longrightarrow Y^m$ be a smooth, proper and surjective map of nonsingular complex quasi-projective varieties of the indicated dimensions and L be a an ample line bundle on X. Then

$$L^i : R^{n-m-i} f_* \mathbb{Q}_X \simeq R^{n-m+i} f_* \mathbb{Q}_X, \qquad \forall i \geq 0,$$

$$R f_* \mathbb{Q}_X \simeq \bigoplus_{i \geq 0} R^j f_* \mathbb{Q}_X [-j] \tag{8.2}$$

and the local systems $R^j f_* \mathbb{Q}_X$ are semi-simple on Y.

Proof. See [Deligne, 1969] and [Deligne, 1971], Théorème 4.2.6.   □

Remark 8.2.5. (8.2) implies the E_2-degeneration of the Leray spectral sequence for maps as in Theorem 8.2.4. This is another special feature of complex algebraic geometry.

8.3   The Leray spectral sequence

Theorem 8.3.1. *(The Leray spectral sequence) Let $f : X \to Y$ be a continuous map of topological spaces and G be a sheaf of abelian groups on X.*

There is a first quadrant spectral sequence

$$E_2^{p,q} = H^p(Y, R^q f_* G) \implies H^{p+q}(X, G).$$

Proof. See [Griffiths and Harris, 1978], [Bott and Tu, 1986], or [Demailly, http].   □

Let us spell out the content of Theorem 8.3.1.

For every pair of indices (p, q), the abelian group $H^{p+q}(X, G)$ admits a decreasing filtration associated with the map f:

$$H^{p+q}(X, G) = F^0 \supseteq F^1 \supseteq \ldots \supseteq F^{p+q} \supseteq \{0\}.$$

There is a collection of abelian groups $E_r^{p,q}$ and of group homomorphisms $d_r^{p,q}$, $p, q \in \mathbb{Z}$, $r \geq 2$ such that:

– $E_r^{p,q} = \{0\}$ as soon as either $p < 0$ or $q < 0$ (this is what "first quadrant" means);

– $E_2^{p,q} = H^p(Y, R^q f_* G)$;

– $d_r^{p,q} : E_r^{p,q} \longrightarrow E_r^{p+r,q-r+1}$; note that for any fixed $p + q$, $d_r^{p,q} = 0$ for every $r \gg 0$;

– $d_r^2 = 0$;

– $E_{r+1}^{p,q} = \operatorname{Ker} d_r^{p,q} / \operatorname{Im} d_r^{p-r,q+r-1}$; note that for any fixed $p+q$, $E_r^{p,q} = E_{r+1}^{p,q}$ for every $r \gg 0$; we denote these stabilized groups by $E_\infty^{p,q}$;

– $F^p / F^{p+1} = E_\infty^{p,q}$.

One says that the spectral sequence degenerates at E_r if $d_{r'} = 0$ for every $r' \geq r$.

Exercise 8.3.2. Show that there are canonical injective and surjective maps

$$H^{p+q}(X, G) \longrightarrow\!\!\!\!\!\gg F^0/F^1 = E_\infty^{0,p+q} \hookrightarrow E_r^{0,p+q},$$
$$E_r^{p+q,0} \longrightarrow\!\!\!\!\!\gg E_\infty^{p+q} = F^{p+q}/\{0\} \hookrightarrow H^{p+q}(X, G).$$

These maps are called the *edge homomorphism*.

Show that one always has exact sequences

$$0 \longrightarrow H^1(Y, f_*G) \longrightarrow H^1(X, G) \longrightarrow E_\infty^{0,1} \longrightarrow 0,$$
$$0 \longrightarrow H^1(Y, f_*G) \longrightarrow H^1(X, G) \longrightarrow H^0(Y, R^1 f_*G)$$
$$\longrightarrow H^2(Y, f_*G) \longrightarrow H^2(X, G).$$

Exercise 8.3.3. Let $f : S^n \longrightarrow S^m$ be a fiber bundle with fiber S^{n-m}, i.e. a bundle structure on a sphere, over a sphere in spheres. Use the Leray spectral sequence to determine the possible values of (n, m). (Hint: for $m = 1$ use the fundamental groups; for $m \geq 2$ use the fact that S^m being simply connected implies that the local systems $R^j f_* \mathbb{Z}_{S^n}$ are constant (cf. [Bott and Tu, 1986], Theorems II.13.2 and II.13.4). Find examples of such bundles.

8.4   The Global Invariant Cycle Theorem

Let $f : U \to S$ be a proper and smooth morphism of smooth complex algebraic varieties. In particular, the fibers of f are smooth, possibly disconnected, compact smooth complex manifolds and any two fibers of f are diffeomorphic to each other.

As discussed in Example 8.2.1, we have the *monodromy representation*

$$\pi_1(S, s) \longrightarrow Aut(H^j(f^{-1}(s), \mathbb{Q})).$$

Every cohomology class on U restricts to a $\pi_1(S, s)$-invariant class in $H^j(f^{-1}(y), \mathbb{Q})$.

Theorem 8.4.1. *(The Global Invariant Cycle Theorem)* Global Invariant Cycle Theorem *Let $U \subseteq X$ be a smooth compactification of U. Then the natural restriction map*

$$H^j(X, \mathbb{Q}) \longrightarrow H^j(f^{-1}(s), \mathbb{Q})^{\pi_1(S,s)}$$

is surjective, i.e the image is given by the invariants.

Proof. By (8.2), Remark 8.2.5, Exercise 8.3.2 and (8.1), the Leray spectral sequence for f is E_2-degenerate and we get a natural surjection

$$H^j(U,\mathbb{Q}) \longrightarrow H^0(S, R^j f_* \mathbb{Q}_U) \simeq H^j(f^{-1}(s),\mathbb{Q})^{\pi_1(S,s)}.$$

The conclusion follows from Theorem 8.1.2. □

A typical situation to which this theorem applies is when one has a proper surjective morphism $g : X \to Y$ with X smooth. Then one sets $S \subseteq Y$ to be the locus over which g is smooth, $U := g^{-1}(S)$ and $f := g_{|U}$.

Every cohomology class on X restricts to a $\pi_1(S,s)$-invariant class in $H^j(f^{-1}(y),\mathbb{Q})$.

The Global Invariant Cycle Theorem establishes the highly non-trivial fact, false in general in a non-Kähler context, that *every* invariant class on the fiber is the restriction of a global class on X.

8.5 The Lefschetz Theorems and semi-simplicity

We now show that the Weak Lefschetz Theorem and the Semi-simplicity Theorem imply the Hard Lefschetz Theorem.

We need the following simple topological fact which is a consequence of the discussion of §7.2.

Exercise 8.5.1. Let $X \subseteq \mathbb{P}^N$ be a projective manifold, Y be the transverse intersection of X with a general hyperplane, $r : Y \to X$ be the natural map, $L := L_{1|X}$, $L' := L_{|Y}$.

Show that the following diagram is commutative:

$$
\begin{array}{ccc}
H^\bullet(X,\mathbb{Q}) & \xrightarrow{L^j} & H^{\bullet+2j}(X,\mathbb{Q}) \\
\downarrow r^* & & \uparrow \tilde{r}_* \\
H^\bullet(Y,\mathbb{Q}) & \xrightarrow{L'^{j-1}} & H^{\bullet+2j-2}(Y,\mathbb{Q}).
\end{array}
\tag{8.3}
$$

Exercise 8.5.2. Let $f : V \longrightarrow W$ be a linear map of finite dimensional vector spaces, $f^* : W^* \longrightarrow V^*$ the map dual to f, ϕ' a non-degenerate bilinear form on W, $\phi : W \simeq W^*$ the resulting isomorphism and $l := f^* \circ \phi \circ f$:

$$\begin{array}{ccc} V & \xrightarrow{\ l\ } & V^* \\ \downarrow f & & \uparrow f^* \\ W & \xrightarrow{\ \phi\ } & W^*. \end{array}$$

Prove that

$$\operatorname{Ker} f^* \circ \phi = (\operatorname{Im} f)^{\perp_{\phi'}}.$$

Deduce that, given the diagram (8.3) for $\bullet = n - 1$:

$$\begin{array}{ccc} H^{n-1}(X,\mathbb{Q}) & \xrightarrow{\ L\ } & H^{n+1}(X,\mathbb{Q}) \\ \searrow {\scriptstyle r^*} & & \nearrow {\scriptstyle \tilde{r}_*} \\ & H^{n-1}(Y), & \end{array} \qquad (8.4)$$

we have that

$$\operatorname{Ker} \tilde{r}_* = (\operatorname{Im} r^*)^{\perp}.$$

Show that L is an isomorphism iff

$$\operatorname{Ker} \tilde{r}_* \cap \operatorname{Im} r^* = \{0\}$$

iff the restriction of the intersection form on $H^{n-1}(Y,\mathbb{Q})$ to the injective image of

$$H^{n-1}(X,\mathbb{Q}) \longrightarrow H^{n-1}(Y,\mathbb{Q})$$

is non-degenerate.

We also need the following elementary algebraic fact.

Exercise 8.5.3. Let V be a finite dimensional representation of a group π, ψ a π-invariant bilinear form on V, $V^\pi \subseteq V$ be the subspace of π-invariant vectors.

Assume that V is completely reducible, i.e. that every π-invariant subspace V' of V admits a complement V'', i.e. a π-invariant subspace V'' such that $V = V' \oplus V''$.

Show that if ψ is non-degenerate, then $\psi_{|V^\pi}$ is non-degenerate. (Hint: consider a complement W of V^π, assume that V^π and W are not orthogonal; deduce that W admits a π-invariant subspace of dimension 1 on which π acts trivially; this leads to a contradiction.)

Let s be the section of L giving the smooth $Y \subseteq X$, s' another general section whose smooth zero-locus Y' meets Y transversally at $Z := Y \cap Y'$. Let $p : \widetilde{X} \longrightarrow X$ be the blowing up of X along Z, $q : \widetilde{X} \longrightarrow \mathbb{P}^1$ be the resulting map associated with the pencil of hyperplane sections associated with s and s'. We have a canonical embedding $Y \subseteq \widetilde{X}$ as the fiber of q over the point, still called s, of the pencil corresponding to s.

Exercise 8.5.4. Verify that the images of $H^j(\widetilde{X}, \mathbb{Q})$ and of $H^j(X, \mathbb{Q})$ in $H^j(Y, \mathbb{Q})$ coincide.

Let $U \subseteq \mathbb{P}^1$ be the locus over which q is smooth. Note that $s \in U$.

After all these preliminaries we can now draw the bridge between the Weak and the Hard Lefschetz Theorem.

Proposition 8.5.5. *Assume that the Hard Lefschetz Theorem has been proved for Y.*

Then the Hard Lefschetz Theorem for X is equivalent to the statement that the intersection form on $H^{n-1}(Y, \mathbb{Q})$ is non-degenerate when restricted to the injective image of $H^{n-1}(X, \mathbb{Q})$ in $H^{n-1}(Y, \mathbb{Q})$.

This last statement, in turn, is implied either by (a) the Semi-simplicity Theorem, or by (b) the Hodge-Riemann Bilinear Relations on Y.

Proof. We specialize the diagram (8.3) to

$$
\begin{array}{ccc}
H^{n-j}(X, \mathbb{Q}) & \xrightarrow{L^j} & H^{n+j}(X, \mathbb{Q}) \\
\downarrow r^* & & \uparrow \widetilde{r}_* \\
H^{(n-1)-(j-1)}(Y, \mathbb{Q}) & \xrightarrow{L'^{j-1}} & H^{(n-1)+(j-1)}(Y, \mathbb{Q})
\end{array}
$$

for $j = 0$ and $j \geq 2$.

The Weak Lefschetz Theorem and the hypotheses imply that L^j is an isomorphism for $j = 0$ and for $j \geq 2$.

The critical case is $j = 1$ which gives the diagram (8.4).

By taking a pencil as above, applying the Global Invariant Cycle Theorem 8.4.1 and Exercise 8.5.4 we deduce that

$$
H^{n-1}(X, \mathbb{Q}) = H^{n-1}(Y, \mathbb{Q})^{\pi_1(U,s)}.
$$

By Exercise 8.5.2, we are left with proving that the restriction of the intersection form on $H^{n-1}(Y, \mathbb{Q})$ to (the injective image of) $H^{n-1}(X, \mathbb{Q})$ is non-degenerate.

The Semi-simplicity Theorem, i.e. condition (a), implies that $H^{n-1}(Y, \mathbb{Q})$ is a completely reducible $\pi_1(U, s)$-representation.

The sought-for non-degeneration follows from Exercise 8.5.3. So that the Hard Lefschetz Theorem for X follows using (a).

The Primitive Lefschetz Decomposition on Y gives

$$H^{n-1}(Y, \mathbb{Q}) = P_{L'}^{n-1} \overset{\perp}{\oplus} L' H^{n-3}(Y, \mathbb{Q}).$$

Using the Weak Lefschetz Theorem we obtain

$$H^{n-1}(X) = \left(P_{L'}^{n-1} \cap H^{n-1}(X) \right) \overset{\perp}{\oplus} L' H^{n-3}(Y, \mathbb{Q}).$$

Since the intersection form on $H^{n-1}(Y, \mathbb{Q})$ is non-degenerate by Poincaré Duality, it is non-degenerate on the direct summand $L' H^{n-3}(Y, \mathbb{Q})$. By orthogonality the sought-for non-degeneration is equivalent to the intersection form being non-degenerate on the space $P_{L'}^{n-1} \cap H^{n-1}(X)$.

This space is a sub-Hodge structure of the pure Hodge structure $P_{L'}^{n-1} \subseteq H^{n-1}(Y, \mathbb{Q})$ which is polarized by the intersection form by virtue of the Hodge-Riemann Bilinear relations on Y.

By Remark 7.1.9 this polarization restricts to a polarization of the Hodge sub-structure $P_{L'}^{n-1} \cap H^{n-1}(X)$ so that the intersection form is non-degenerate on this space; see Remark 7.1.8. $\qquad\square$

Proposition 8.5.5 does *not* yield a proof of the Hard Lefschetz Theorem, which is a topological statement concerning the cup product operation with the fundamental class of a hyperplane section, independently of Hodge Theory.

In fact, in both cases (a) and (b), the known transcendental ways to prove that these assumptions are met is through Hodge theory, in fact through the Hard Lefschetz Theorem!

In his second paper on the Weil Conjectures, Deligne has given an algebraic proof of the Hard Lefschetz Theorem for the l-adic cohomology of a nonsingular projective varieties over algebraically closed fields of positive characteristic.

The result implies the Hard Lefschetz Theorem 7.3.4.a. However, presently, there is no "algebraic" proof of this result that does not make use of algebraic geometry in positive characteristic.

8.6 Approximability for the space of primitive vectors

In this final section I would like to discuss, in a simple case, two of the techniques introduced in the papers [de Cataldo and Migliorini, 2002] and [de Cataldo and Migliorini, 2005]. These papers deal with the new structures on the rational singular cohomology of a projective manifold that arise in connection with a projective map $f : X \longrightarrow Y$.

The set-up of this section is as follows.

Let $f : X \to Y$ be a map of projective varieties, X be nonsingular, L be a hyperplane bundle on X, H be a hyperplane bundle on Y, $M := f^*H$.

The first technique is used to deal, in some cases, with the natural class map

$$H_\bullet(f^{-1}(y)) \longrightarrow H^{2n-\bullet}(X), \qquad y \in Y.$$

The second is an approximation technique for certain cohomology classes which are "primitive" with respect to a line bundle M which is generated by its global sections, but that is not necessarily a hyperplane bundle, on a projective manifold using cohomology classes which are primitive with respect to certain related hyperplane line bundles.

The upshot is that via this approximation it is possible to study some of the properties of the intersection form on these "primitive" spaces.

In order to exemplify these techniques, we discuss a very special, yet meaningful case:

Theorem 8.6.1. *(Contractibility criterion) Let $f : X \to Y$ be a map of projective varieties of even complex dimension $2m$, X be nonsingular, $y \in Y$ be such that $f^{-1}(y)$ is a union of finitely many algebraic varieties E_j in X of dimension m and f is an isomorphism over $Y \setminus y$.*
Then the intersection matrix

$$(-1)^m \|E_j \cdot E_k\|$$

is symmetric positive definite.

Remark 8.6.2. Let us remark that the line bundle $M := f^*H$ is trivial on the varieties E_j so that it cannot be ample. In particular, a priori one cannot apply to it the conclusions of the Hard Lefschetz Theorem and of the classical Hodge-Riemann Bilinear Relations.

There is a re-formulation of Theorem 8.6.1 that holds for *every* projective map with domain an algebraic manifold. The precise statement requires substantial preparation and is omitted. See [de Cataldo and Migliorini, 2002] and [de Cataldo and Migliorini, 2005].

The case $m = 1$ in Theorem 8.6.1 is classical and is the celebrated Grauert criterion for the contractibility of curves on surfaces.

For $m \geq 2$, Theorem 8.6.1 seems new. See [de Cataldo and Migliorini, 2002] and [de Cataldo and Migliorini, 2005]. For example, if a morphism $f : X \to Y$ contracts a surface E in a fourfold X and nothing else, then $[E]^2 > 0$.

Example 8.6.3. As a toy-model, the reader can keep in mind the special case $m = 1$ of Theorem 8.6.1 in all the considerations that follow.

Proposition 8.6.4. *(Linear independence of the fundamental classes) Let things be as in Theorem 8.6.1. Then the fundamental classes $[E_j] \in H^{2m}(X, \mathbb{Q})$ of the contracted varieties are linearly independent.*

In particular, the natural class map $H_{2m}(f^{-1}(y), \mathbb{Q}) \longrightarrow H^{2m}(X, \mathbb{Q})$ is an injection of pure Hodge structures of weight $2m$.

Proof. Since $H_{2m}(f^{-1}(y), \mathbb{Q})$ is a pure Hodge structure of weight $2m$, spanned by the fundamental homology classes $\{E_j\}$ (which are of type (m, m)) we only need to prove that the class map is injective.

Let U be an affine neighborhood of y in Y, e.g. $U = Y \setminus H$, where H is a hyperplane section of Y, relative to some embedding, not containing y.

Let $U' := f^{-1}(U)$ and $g := f_{|U'} : U' \longrightarrow U$. We look at the Leray spectral sequence for g.

We have natural isomorphisms

$$R^j g_* \mathbb{Q}_{U'} \simeq H^j(g^{-1}(y), \mathbb{Q}), \qquad j > 0, \qquad (8.5)$$

where the right-hand side is viewed as a skyscraper sheaf at y.

The sheaves $R^j g_* \mathbb{Q}_{U'}$ are what one calls *constructible sheaves* on U, for every j. In this context, it only means that they are locally constant when restricted to $U \setminus y$ and to y.

Since U is affine, the theorem on the cohomological dimension of affine sets with respect to constructible sheaves (which we do not state in these lectures, but which is the natural generalization of the Andreotti-Frankel result on the homotopy type of affine complex manifolds quoted in the proof of the Weak Lefschetz Theorem 7.4) implies that

$$H^p(U, R^0 g_* \mathbb{Q}_{U'}) = 0, \qquad p > 2. \tag{8.6}$$

On the other end, (8.5) and (8.6) imply that $d_r^{0,2m} = 0$ for every $r \geq 2$: these differentials land in zero groups.

It follows that $E_\infty^{0,2m} = E_2^{0,2m} \simeq H^{2m}(f^{-1}(y), \mathbb{Q})$.

By Exercise 8.3.2, the surjectivity of the edge map implies that the natural restriction map

$$H^{2m}(U', \mathbb{Q}) \longrightarrow H^{2m}(f^{-1}(y), \mathbb{Q})$$

is surjective.

By Theorem 8.1.2, the restriction map

$$H^{2m}(X, \mathbb{Q}) \longrightarrow H^{2m}(f^{-1}(y), \mathbb{Q})$$

is surjective as well.

The dual map, i.e. the class map,

$$H_{2m}(f^{-1}(y), \mathbb{Q}) \longrightarrow H_{2m}(X, \mathbb{Q}) \overset{PD}{\simeq} H^{2m}(X, \mathbb{Q})$$

is injective. $\qquad\qquad\qquad\qquad\qquad\qquad\qquad\qquad\qquad\quad\square$

Exercise 8.6.5. Let (X, ω) be a Kähler manifold, $f : X \to \mathbb{P}^N$ be a holomorphic map. Show that $f^* \omega_{FS} + \epsilon \omega$ is a Kähler form for every $\epsilon > 0$.

Deduce that if $f : X \to Y$, L, H and M are as in the set-up, then one can represent the real classes $M + \epsilon L := c_1(M) + \epsilon c_1(L) \in H^{1,1}(X)$ using Kähler classes (i.e. associated with Kähler metrics). In particular, the Hard Lefschetz Theorem and the Hodge-Riemann Bilinear Relations hold for $M + \epsilon L$.

Exercise 8.6.6. (**The Hard Lefschetz Theorem for M**; cf. [de Cataldo and Migliorini, 2002]) Show that the Hard Lefschetz Theorem 7.3.4.a holds for M, i.e. show that

$$M^r : H^{2m-r}(X, \mathbb{Q}) \simeq H^{2m+r}(X, \mathbb{Q}).$$

(Hint: First prove the Weak Lefschetz Theorem for the smooth sections Y of M on X as in Theorem 7.4.1; the relevant vanishing can be proved along the lines of the proof of Proposition 8.6.4. Note that one can always choose Y to avoid $f^{-1}(y)$. Use Proposition 8.5.5: the point is to be able to use condition (b) in that theorem on Y which can be done since $M_{|Y}$ is a hyperplane bundle.)

Show that the Primitive Lefschetz Decomposition Theorem 7.3.4.b holds for M, in particular, having set $P_M^{2m} := \operatorname{Ker} M \subseteq H^{2m}(X, \mathbb{Q})$:

$$H^{2m}(X, \mathbb{Q}) = P_M^{2m} \overset{\perp}{\oplus} M\, H^{2m-2}(X, \mathbb{Q}) \tag{8.7}$$

and, recalling Exercise 8.6.5,

$$\dim_\mathbb{Q} P_M^{2m} = b_{2m} - b_{2m-2} = \dim_\mathbb{R} P_{M+\epsilon L}^{2m}(X, \mathbb{R}). \tag{8.8}$$

Proposition 8.6.7. (Approximability of P_M^{2m}) *Let things be as in Theorem 8.6.1.*

Then, in the Grassmannian $G(b_{2m} - b_{2m-2}, H^{2m}(X, \mathbb{R}))$ we have

$$P_M^{2m}(X, \mathbb{R}) = \lim_{\epsilon \to 0} P_{M+\epsilon L}^{2m}(X, \mathbb{R})$$

and

$$\Psi := (-1)^m \int_X - \wedge -$$

defines a polarization of $P_M^{2m}(X, \mathbb{R})$.

Proof. We have an elementary inclusion of vector spaces

$$P_M^{2m}(X, \mathbb{R}) \supseteq \lim_{\epsilon \to 0} P_{M+\epsilon L}^{2m}(X, \mathbb{R}).$$

By (8.8), both sides have the same finite dimension so that they coincide.

The Hodge-Riemann Bilinear Relations for $M + \epsilon L$ imply that Ψ is a polarization for $P_{M+\epsilon L}^{2m}(X, \mathbb{R})$, for every $\epsilon > 0$.

It follows that $\Psi(-, C(-))$ is positive semi-definite on the limit $P_M^{2m}(X, \mathbb{R})$.

By (8.7), since Ψ is non-degenerate on $H^{2m}(X, \mathbb{Q})$, it remains non-degenerate on $P_M^{2m}(X, \mathbb{R})$.

Since C is a real automorphism, $\Psi(-, C(-))$, is also non-degenerate on $P_M^{2m}(X, \mathbb{R})$.

It follows that $\Psi(-, C(-))$, must be positive definite, i.e. that Ψ is a polarization of $P_M^{2m}(X, \mathbb{R})$. $\qquad\square$

Proof of Theorem 8.6.1. Note that since one can choose a section of H avoiding y, the injective image of the class map lands in the M primitive space

$$H_{2m}(f^{-1}(y)) \subseteq P_M^{2m} \subseteq H^{2m}(X, \mathbb{Q})$$

as a sub-Hodge structure.

By Proposition 8.6.7 and Remark 7.1.9, $\Psi = (-1)^m \int_X - \wedge -$ is a polarization for this image so that $\Psi(-, C(-))$ is positive definite on $H_{2m}(f^{-1}(y))$.

The statement follows from the fact that all classes under consideration are rational and of type $(p, q) = (m, m)$ so that C acts as the identity on $H_{2m}(f^{-1}(y))$. $\square$

Bibliography

Bott, R., Tu, L.W.: *Differential Forms in Algebraic Topology*, GTM 82, Springer-Verlag, Berlin Heidelberg New York, Second Printing 1986.

de Cataldo, M.A.A., Migliorini, L.: "The Hard Lefschetz Theorem and the topology of semismall maps," Ann.Scient.Ec.Norm.Sup., 4e serie, t.35. 2002, 759-772.

de Cataldo, M.A.A., Migliorini, L.: "The Hodge theory of algebraic maps," Ann.Scient.Ec.Norm.Sup., 4e serie, t.38. 2005, 693-750.

Deligne, P.: "Théorème de Lefschetz et critères de dégénérescence de suites spectrales," Publ.Math. IHES **35** (1969), 107-126.

Deligne, P.: "Théorie de Hodge, II," Publ.Math. IHES **40** (1971), 5-57.

Deligne, P.: "Théorie de Hodge, III," Publ.Math. IHES **44** (1974), 5-78.

Demailly, J.-P.: "Théorie de Hodge L^2 et théorèmes d'annulation," in *Introduction à la théorie de Hodge*, J. Bertin, J.-P. Demailly, L. Illusie, C. Peters, Panoramas et Synthèses, n.3, Société Mathématique de France, 1996.

Demailly, J.-P.: *Complex Analyitc and Differential Geometry*, available at http://www-fourier.ujf-grenoble.fr/~demailly/books.html

Durfee, A.: "A naive guide to mixed Hodge theory." Singularities, Part 1 (Arcata, Calif., 1981), 313–320, Proc. Sympos. Pure Math., 40.

Griffiths, P., Harris, J.: *Principles of Algebraic Geometry*, John Wiley ans Sons, 1978.

Mok, N.: *Metric rigidity theorems on Hermitian symmetric manifolds*, Series in Pure Mathematics, Vol. 6, World Scientific, 1989.

Mumford, D.: *Algebraic Geometry I, Complex Projective Varieties*, Springer-Verlag, Berlin Heidelberg New York, 1970.

Voisin, C.: *Théorie de Hodge et géométrie algébrique complexe*, Cours Spécialisés **10**, Soc.Math. de France, 2002.

Warner, F. W.: *Foundations of differentiable manifolds and Lie groups*, Scott, Foresman and Company, 1971.

Weil, A.: *Introduction à l' étude des variétés kähleriennes*, Hermann, Paris, 1958.

Wells, R.O.: *Differential analysis on complex manifolds,* Graduate Texts in Math. **65**, 2nd edition, Springer-Verlag, Berlin, 1980.

Index